The Anthropocene: Politik—Economics—Society—Science

Volume 4

Series editor

Hans Günter Brauch, Mosbach, Germany

More information about this series at http://www.springer.com/series/15232
http://www.afes-press-books.de/html/APESS.htm
http://www.afes-press-books.de/html/APESS_04-05.htm

Hans Günter Brauch · Úrsula Oswald Spring
Juliet Bennett · Serena Eréndira Serrano Oswald
Editors

Addressing Global Environmental Challenges from a Peace Ecology Perspective

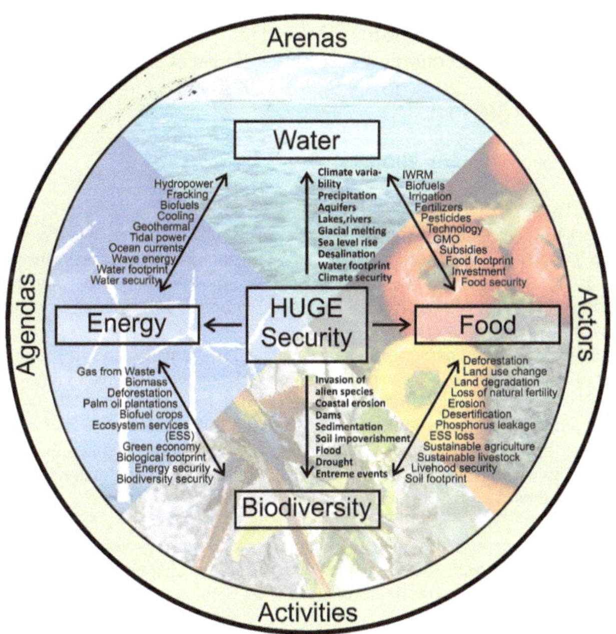

Editors
Hans Günter Brauch
Peace Research and European Security Studies (AFES-PRESS)
Mosbach, Baden-Württemberg
Germany

Juliet Bennett
Center for Peace and Conflict Studies
University of Sydney
Sydney, New South Wales
Australia

Úrsula Oswald Spring
Center for Regional Multidisciplinary Studies (CRIM)
National Autonomous University of Mexico (UNAM)
Cuernavaca, Morelos
Mexico

Serena Eréndira Serrano Oswald
Center for Regional Multidisciplinary Studies (CRIM)
National Autonomous University of Mexico (UNAM)
Cuernavaca, Morelos
Mexico

ISSN 2367-4024 ISSN 2367-4032 (electronic)
The Anthropocene: Politik—Economics—Society—Science
ISBN 978-3-319-30989-7 ISBN 978-3-319-30990-3 (eBook)
DOI 10.1007/978-3-319-30990-3

Library of Congress Control Number: 2016946014

Cover illustration: The cover was designed by Úrsula Oswald Spring with technical assistance from Angel Paredes Rivera, Cuernavaca, Mexico and is taken from her chapter in this volume. More on this book is at: http://www.afes-press-books.de/html/APESS_04-05.htm.

Copyediting: PD Dr. Hans Günter Brauch, AFES-PRESS e.V., Mosbach, Germany
Language Editing: Mike Headon, Colwyn Bay, Wales, Great Britain (Chapters 1, 2, 6, 7)

Printed on acid-free paper

This Springer imprint is published by Springer Nature
The registered company is Springer International Publishing AG Switzerland

Foreword

While the USSR and socialist ideology were considered as the greatest security threat in a polarized international system, particularly for the industrialized Western countries, military security was prioritized and all subjects which were unrelated to military power were seen as secondary, and other causes of conflicts overlooked. However, minority issues, ethnic and religious conflicts, international terrorism, mass migration, the refugee problem, climate change and other environmental issues which emerged after the Cold War were seen as serious threats to international peace and security. After the ideological threat was removed with the end of the Cold War, a turning point was reached in security studies and 'the environment' began to appear in security studies as security was reconceptualized and security threats were redefined. Since then social inequality, population growth, migration and refugee movements, wars waged to protect environmental resources, and the targeting of the natural environment and resources or using them for military purposes have been regarded as social risks related to environmental security.

Environmental security is closely associated with other security issues, and the outcomes caused by environmental issues have negative impacts on other security areas. It has become evident that environmental factors are among the causes of many ethnic, religious and ideological conflicts, and social tensions worldwide.

Thus, such environmental problems as food shortage, water and land shortages, access to natural resources, and destruction of oil and fishery resources have negative impacts on social life, cause disputes between countries, and lead to sharing problems, conflicts or other forms of fighting among countries deprived of resources and countries/groups which own such resources; the environmental factors which directly affect political and economic fields might render migrations unavoidable, and migrations may cause tensions, conflicts or resource competition between residents and new arrivals.

Conflicts in Rwanda and Burundi based on renewable resources such as water, food and forests exemplify this. As a consequence of the growing population and resource depletion in Rwanda, mass killings were carried out by both Hutu and Tutsi peoples. People in Honduras and El Salvador scraping a living off their land were involved in conflicts, and a refugee issue emerged as a result of these conflicts. Environmental disputes also arose in the Democratic Republic of the Congo, Nigeria, the Philippines, Nepal and southern Africa. Some claim that Israel's control of the West Bank is rooted in its desire to control water resources in the region alongside Palestine's hostile attitude. In 2004 during the migration movements in Chad, the United Nations High Commissioner for Refugees noted increased conflicts between refugees and local people, caused by limited resources.

The book edited by Hans Günter Brauch, Úrsula Oswald Spring, Juliet Bennett, and Serena Eréndira Serrano Oswald makes a substantial contribution to the literature. Its chapters were presented at the sessions of the *Ecology and Peace Commission* (EPC) at the 25th conference of the *International Peace Research Association* (IPRA) in İstanbul on the occasion of the centenary of the outbreak of World War I and IPRA's fiftieth anniversary.

Environment, Security, Development and Peace offers innovative research on environmental issues in the social sciences and humanities.

The chapters have been written by authors from Australia, Canada, Finland, Germany and Mexico and from the fields of medicine, clinical psychology, anthropology, ecology, classical and modern languages, anthropology, sociology, gender studies, and political sciences, and adopt a variety of theoretical and methodological approaches.

Chapters address global environmental challenges from a peace ecology perspective and deal with the global ecological crisis as a form of structural violence, thus linking conceptually Johan Galtung's 'structural theory of imperialism' and Alfred Kahn's 'tyranny of small decisions', "The Emotional Dimensions of Ecological Peacebuilding", "Drowning in complexity? Preliminary findings on addressing gender, peacebuilding and climate change in Honduras", and "The Water, Energy, Food and Biodiversity Nexus: New Security Issues in the Case of Mexico".

Hans Günter Brauch discusses historical times and events, changing global contexts, political turning points, and global transformations and transitions which affected security approaches and led to global environmental challenges in the twentieth century between 1914 and 1989/1990.

Nesrin Kenar and Ibrahim Seaga Shaw, Secretaries General of IPRA, are grateful to the distinguished conveners of the Ecology and Peace Commission, Prof. Dr. Úrsula Oswald Spring and Adj. Prof. Dr. Hans Günter Brauch, for their contributions to the IPRA 2014 Conference and for the organization of successful sessions of the Ecology and Peace Commission. Nesrin Kenar and Ibrahim Shaw further thank the editors Hans Günter Brauch, Úrsula Oswald Spring, Juliet Bennett and Serena Eréndira Serrano Oswald, who compiled the papers presented in these sessions, as well as the authors of the chapters in the book and all the reviewers for their constructive comments.[1]

February 2016

Nesrin Kenar, PhD
Assistant Professor in the Department
of International Relations,
Faculty of Political Sciences,
Sakarya University, Turkey,
Secretary General,
International Peace Research Association

[1]Dr. Nesrin Kenar is currently Assistant Professor in the Department of International Relations, Faculty of Political Sciences, Sakarya University, Turkey. She has contributed analyses on globalization, religion and politics, on ethnic conflicts, on the conflicts in former Yugoslavia, on the civil war in Bosnia-Herzegovina, and on international and foreign policy strategies of states. She is the institutional coordinator of European Union projects at Sakarya University. She was President of the *European Peace Research Association* (EUPRA) from 2008 to 2012, as well as a member of the editorial board of *Global Peace International Journal.* She is the author of *The National and International Dimensions of the Yugoslavia Question* (2005). Dr. Kenar has published on international relations, foreign policy, and peace and conflict studies. She holds a PhD from the Department of International Relations in the Institute of Social Science at Marmara University, İstanbul, Turkey.

Acknowledgements

This book and a second volume on *Regional Ecological Challenges for Peace in Africa, the Middle East, Latin America and Asia Pacific* emerged from written papers that were orally presented in the several sessions of the *Ecology and Peace Commission* (EPC) during the 25th Conference of the *International Peace Research Association* (IPRA) in Istanbul in 10–15 August 2014 on the occasion of the 50th anniversary of IPRA and 100 years after the start of World War I on 28 July 1914.

The editors are grateful to Dr. Nesrin Kenar—with Dr. Ibrahim Shaw, co-secretary-general of IPRA (2012–2016)—who organized the Istanbul conference with her able team from Sakarya University at the Bosporus where Europe and Asia meet. We also thank all the sponsors—including the IPRA Foundation—who supported the participation of a few colleagues from developing and low-income countries who had submitted written papers that were assessed with regard to their scientific quality by the two EPC co-organizers as a precondition for their grant.

The four co-editors of these two books would like to thank all the authors who passed the double-blind anonymous peer-review process and subsequently revised their papers taking the many critical comments and suggestions of these reviewers into account. Each chapter was at least reviewed by three external reviewers who are unrelated to the editors and the authors and in most cases also came from different countries.

We would like to thank all the reviewers who spent much time to read and comment on the submitted texts and made detailed perceptive and critical remarks and suggestions for improvements—even for texts that could not be included in both volumes. The texts by the editors had to pass the same review process based on the same criteria. The goal of the editors has been thus to enhance the quality of the submitted texts. The editors were bound by these reviewers' reports, even if they did not necessarily agree with all their comments and decisions on acceptance or rejection.

The following colleagues (in alphabetical order) contributed anonymous reviews:

- Dr. Kwesi Aning, Kofi Annan Centre, Accra, Ghana
- Ms. Juliet Bennett, University of Sydney, Australia
- Prof. Dr. Sigurd Bergmann, Norwegian University of Science and Technology, Trondheim (NTNU), Norway
- Dr. Katharina Bitzker, University of Manitoba, Canada
- Dr. Lynda-Ann Blanchard, University of Sydney, Australia
- Prof. Dr. Michael Bothe, Emeritus, Johann Wolfgang Goethe University, Frankfurt on Main, Germany
- PD Dr. Hans Günter Brauch, ret., Free University of Berlin, Germany
- Mr. Christopher Brown, University of Sydney, Australia
- Dr. Carl Bruch, Environmental Law Institute, Washington, D.C., USA
- Prof. Dr. Halvard Buhaug, Norwegian University of Science and Technology, Trondheim (NTNU), Norway
- Prof. Dr. Ken Conca, American University, Washington, D.C., USA
- Prof. Dr. Hendrix Cullen, University of Denver, Colorado, USA
- Prof. Dr. Paul Custler, Lenoir Rhyne University, North Carolina, USA
- Prof. Dr. Simon Dalby, CIGI Chair, Political Economy of Climate Change, Balsillie School of International Affairs; Professor of Geography and Environmental Studies, Wilfrid Laurier University, Waterloo, Ontario, Canada
- Dr. Paul Duffill, University of Sydney, Australia
- Dr. Josh Fisher, Columbia University, USA
- Dr. Giovanna Gioli, Hamburg University, Germany
- Mr. Karlson Hargroves, Adelaide University, Australia
- Fredrik S. Heffermehl, Nobel Peace Prize Watch, Oslo, Norway
- Dr. Francis Hutchinson, University of Sydney, Australia
- Dr. Tobias Ide, Georg Eckert Institute for School Book Research, Braunschweig, Germany
- Dr. Anders Jägerskop, Stockholm International Water Institute, Stockholm, Sweden
- Dr. Peter King, University of Sydney, Australia
- Jesús Antonio Machuca R., Faculty of Political and Social Sciences, UNAM, Mexico
- Dr. Eyal Mayroz, University of Sydney, Australia
- Dr. Annabel McGoldrick, University of Sydney, Australia
- Prof. Dr. Syed Sikander Mehdi, University of Karachi, Pakistan
- Dr. Bonaventure Mkandawire, Director of Research and Training at Church and Society Programme, CCAP Livingstonia Synod, Mzuzu, Malawi
- Prof. Dr. Michael Northcott, University of Edinburgh, Scotland, UK
- Prof. Dr. Úrsula Oswald Spring, UNAM, CRIM

- Dr. John Pokoo, Kofi Annan Centre, Accra, Ghana
- Prof. Dr. Mary Louise Pratt, New York University, New York, USA
- Abe Quadan, University of Sydney, Australia
- Prof. Dr. Daniel Reichman, University of Rochester, New Jersey, USA
- Prof. Dr. Luc Reychler, Emeritus, University of Leuven, Belgium; former Secretary-General of IPRA
- Dr. Vivianna Rodriguez Carreon, University of Sydney, Australia
- Dr. Hilmi Salem, Bethlehem, Palestine
- Prof. Dr. Salvany Santiago, Federal University of Sao Francisco Valley, Brazil
- Dr. Janpeter Schilling, Hamburg University, Germany
- Dr. Klaus Schlichtmann, Nihon University, Japan
- Dr. Ayesha Siddiqi, King's College London, London, UK
- Dr. Sunil Tankha, Institute for Social Studies (ISS), The Hague, The Netherlands
- Andres Macias Tolosa, Universidad Externado de Colombia, Colombia
- Em. Prof. Dr. Garry Trompf, University of Sydney, Australia
- Prof. Dr. Thanh-Dam Truong, Emerita, Institute for Social Studies (ISS), The Hague, The Netherlands
- Prof. Dr. Catherine M. Tucker, Indiana University, Indiana, USA
- Prof. Dr. Arthur H. Westing, Westing Associates, Vermont, USA; ret., SIPRI, PRIO
- Prof. Dr. Kazuyo Yamane, Ritsumeikan University, Kyoto, Japan

These two books are the result of an international teamwork among the editors and convenors of IPRA's EPC. As co-convenors, *Prof. Dr. Úrsula Oswald Spring* and *PD Dr. Hans Günter Brauch* organized several sessions of IPRA's EPC in Istanbul and are also the two lead authors of the introductory chapters of both books. Hans Günter Brauch prepared both volumes, managed the peer-review process and did the copy-editing. As a native English speaker, *Ms. Juliet Bennett* (Sydney University, Australia)—who was elected in Istanbul as the third EPC co-convener—language-edited the contributions of the second book and also wrote the concluding essay of the second volume.

The publication and production of this book was handled by an able female team of editors and producers at Springer's office in Heidelberg coordinated by *Dr. Johanna Schwarz*, senior publishing editor, focused on earth system sciences, marine geosciences, palaeoclimatology, polar sciences and volcanology, and *Janet Sterritt-Brunner* (producer and project coordinator) both working at Springer's editorial office in Heidelberg, Germany, and *Ms. Divya Selvaraj* who coordinated the typesetting and production of the book in Chennai, Tamil Nadu, India. Thus, this book is the result of a close cooperation among authors, reviewers and producers from all five continents. The editors are looking forward to see both new readers, speakers and authors at IPRA's next conference in Freetown (Sierra Leone) in November/December 2016.

Mosbach, Germany — Hans Günter Brauch
Cuernavaca, Mexico — Úrsula Oswald Spring
Sydney, Australia — Juliet Bennett
Cuernavaca, Mexico — Serena Eréndira Serrano Oswald
December 2015

Contents

Abbreviations

AFES-PRESS	Peace Research and European Security Studies
AGW	Anthropogenic global warming
ANZUS	Australian, New Zealand and US (military alliance)
AR5	Fifth scientific assessment
ASEAN	Association of South East Asian Nations
BBC	British Broadcasting Corporation
BP	Before present
BRICS	Brazil, Russia, India, China, South Africa
BS	Biodiversity security
CBD	Convention on Biological Diversity of the United Nations
CC	Climate change
CDM	Clean development mechanism
CEM	Clean Energy Ministerial
CEOs	Chief Executive Officer
CH_4	Methane
CO_2	Carbon dioxide
COP	Conference of Parties
COP15	Conference of Parties of UNFCCC in Copenhagen in 2009
COP21	Conference of Parties of UNFCCC in Paris in December 2015
CRIM	Center for Multidisciplinary Studies (of UNAM, in Mexico)
CSCE	Conference for Security and Cooperation in Europe, since 1992 OSCE
CSD	United Nations' Commission on Sustainable Development
DIVERSITAS	International research programme on biodiversity science
EDGAR	Database of European Commission, Netherlands Environmental Assessment Agency
ENS	Energy security
EPC	Ecology and Peace Commission

EPICA Dome C	European Project for Ice Coring in Antarctica
ES	Environmental security
ESS	Ecosystem services
ESSP	Earth Systems Science Programme
EU	European Union
FAO	Food and Agriculture Organization (of the United Nations)
FS	Food security
FY	Fiscal year
G-7	Group of 7 large industrial countries (Canada, France, Germany, Italy, Japan, UK, USA and European Union)
G-20	Group of 20 major industrialized and industrializing countries
GDP	Gross domestic product
GEC	Global environmental change
GHG	Greenhouse gases
GISP 2	Greenland Ice Sheet Project
GMO	Genetically modified organisms
GSA	Geological Survey of America
GTS	Geologic time
GW	Gigawatts
ha	Hectares
HS	Human security
HUGE	Human, gender and environmental (security)
IASA	The International Institute for Applied Systems Analysis
ICS	International Commission on Stratigraphy
IEA	International Energy Agency
IGBP	International Geosphere–Biosphere Programme
IHDI	Inequality-adjusted Human Development Index
IHDP	International Human Dimensions Programme on Global Environmental Change
IMF	International Monetary Fund
IPCC	Intergovernmental Panel on Climate Change
IPRA	International Peace Research Association
IR	International relations
IS	Islamic State (or so-called Kalifad in parts of Syria and Iraq)
ISS	Institute for Social Studies
IWRM	Integrated Water Resources Management
KP	Kyoto Protocol
KSI	Dutch Knowledge Network on Systems Innovation and Transition
LED	Light-emitting diode
LULUCF	Land use, land-use change and forestry
MCFD	Million cubic feet per day
Mdb	Millions of barrels per day

MVMC	Metropolitan Valley of Mexico City
N_2O	Nitrous oxide
NAFTA	North American Free Trade Agreement
NATO	North Atlantic Treaty Organization
NGO	Non-governmental Organization
NIC	US National Intelligence Council (CIA)
NOAA	US National Oceanic and Atmospheric Administration
NSC-68	(US) National Security Council (paper No. 68 of January 1950)
NTNU	Norwegian National Technical University Trondheim, Norway
OAS	Organization of American States
OECD	Organization for Economic Co-operation and Development
OEEC	Organization for European Economic Cooperation
OSCE	Organization for Security and Cooperation in Europe
PAN	Partido Acción Nacional (Party of National Action), Mexico
PD	German academic title that is awarded after the habilitation (2nd PhD)
PEISOR	Pressure–effects–impacts–societal–response
PhD	Philosophical Doctorate
PNAS	Proceedings of the National Academy of Sciences (of the USA)
PRI	Partido Revolucionario Institucional (Party of Institutional Revolution), Mexico
PRIO	International Peace Research Institute in Oslo (Norway)
RCP2.6[4.5/6.0/8.5]	Representative Concentration Pathways (RCPs) adopted by the IPCC for AR5
SD	Sustainable development
SDGs	Sustainable Development Goals
SEATO	Southeast Asia Treaty Organization
SGBV	Sexual and gender-based violence
SIPRI	Stockholm International Peace Research Institute
STRN	Sustainability Transition Research Network
TCC	Transnational capitalist class
UCDP	Uppsala Conflict Data Program
UK	United Kingdom
UN	United Nations
UNAM	Universidad Nacional Autónoma de México (National Autonomous University of Mexico)
UNCCD	United Nations Convention to Combat Desertification
UNCED	United Nations Conference on Environment and Development
UNCSD	United Nations Conference on Sustainable Development (Rio+20)

UNDP	United Nations Development Programme
UNEP	United Nations Environment Programme
UNFCCC	United Nations Framework Convention on Climate Change
UNGA	United Nations General Assembly
UNSC	United Nations Security Council
UNSG	United Nations Secretary-General
US	Unites States (of America)
USA	United States of America
US EPA	US Environment Protection Agency
US-EIA	US Energy Information Administration
USSR	Union of Socialist Soviet Republics
WBG	Wissenschaftliche Buchgesellschaft
WBGU	German Advisory Council on Global Change
WCED	World Commission on Environment and Development
WCRP	World Climate Research Programme
WEF	World Economic Forum
WEF&B	Water, energy, food and biodiversity
WMO	World Meteorological Organization
WS	Water security
WWF	World Wide Fund for Nature

Chapter 1 Introduction: Addressing Global Environmental Challenges from a Peace Ecology Perspective

Hans Günter Brauch, Úrsula Oswald Spring, Juliet Bennett and Serena Eréndira Serrano Oswald

1.1 Peace Ecology in the Anthropocene

This is the first of two volumes based on peer reviewed and thoroughly revised scientific presentations, most of them initially discussed during the sessions of the *Ecology and Peace Commission* (EPC) at the *International Peace Research Association* (IPRA) 50th Anniversary Conference in Istanbul in August 2014. This introduction builds on the previous volume (Oswald Spring et al. 2014a, b) in which authors conceptualized peace and ecology, drew multiple linkages between the concepts, and offered initial thoughts on an emerging 'peace ecology' concept first suggested by Kyrou (2007a, b) and recently discussed by Amster (2015). The need for dialogue between environmental and peace studies was first addressed by Boulding (1966), a renowned economist, and his wife Elise Boulding, a peace education specialist and former Secretary-General of IPRA (1988–1990) during the peaceful global 'turn' that ended the Cold War.

The co-editors built on the emerging concept of the Anthropocene (Crutzen/Stoermer 2000; Crutzen 2002, 2011), a new era in earth and human history recognized by humankind's dominant influence on nature's systems. Since

PD Dr. Hans Günter Brauch, chairman, Peace Research and European Security Studies (AFES-PRESS), since 1987; co-convenor, IPRA's Ecology and Peace Commission (2012–2016), Mosbach, Germany; Email: brauch@afes-press.de.

Prof. Dr. Úrsula Oswald Spring, full-time Professor/Researcher at the *National University of Mexico* (UNAM) in the *Regional Multidisciplinary Research Center* (CRIM); co-convenor, IPRA's Ecology and Peace Commission (2012–2016); Email: uoswald@gmail.com.

Juliet Bennett, Ph.D. candidate, University of Sydney; co-convenor, IPRA's Ecology and Peace Commission (2014–2016); Email: juliet.bennett@sydney.edu.au.

Prof. Dr. Serena Eréndira Serrano Oswald; research-professor, *Regional Multidisciplinary Research Center* (CRIM); president of AMECIDER Mexico; Email: sesohi@hotmail.com.

H.G. Brauch et al. (eds.), *Addressing Global Environmental Challenges from a Peace Ecology Perspective*, The Anthropocene: Politik—Economics—Society—Science 4, DOI 10.1007/978-3-319-30990-3_1

the industrial revolution and the end of World War II, humanity's energy consumption has triggered climatic impacts and dangerous socio-political outcomes that must be addressed in a proactive mode by societal, economic and political actors and scientists. Since the early 1970s, natural scientists have gradually 'socially constructed' issues of *global environmental change* (GEC) and *climate change* (CC), putting these new challenges on the political agenda during the 1980s where they became issues of global diplomacy.

Neither Kyrou nor Amster have offered scientific definitions of 'peace ecology' as a new scientific research programme. For Amster (2015: 20) peace ecology "encompasses the pragmatics of creating and sustaining human societies in their material as well as their ideological needs." Amster (2015: 204) sees "peace among ourselves [as] contingent upon and necessarily related to our ability to live peacefully on earth". He concluded that:

> Peace ecology is eminently cyclical, fostering interconnections and generating dynamic exchange, illuminating the synergies between self, society and nature, and seeking to connect the past, present and future… We have posited that a working version of 'structural peace' could serve as a constructive counterpoint to 'structural violence' … and that a vision of interconnectedness may be the antidote to apathy, isolation and despair.

Thus, peace ecology is here being conceived primarily as a 'political concept' within an 'action perspective,' and not as a scientific concept and research paradigm or programme. For the co-editors of this book, 'peace ecology in the Anthropocene' refers to the goal of 'peace' (in its multiple dimensions as positive, negative, cultural, engendered and sustainable peace) from the perspective of 'ecology'. Ecology has expanded its meaning from the biophysical sciences after World War II, to include the social sciences and humanities. Peace ecology in the Anthropocene aims to address human-induced changes in the earth system, and lead them toward peaceful alternatives (Oswald Spring et al. 2014b).

While Dalby (2013a, b, 2014, 2015) has discussed conceptual issues of security during the Anthropocene, this chapter approaches the socio-political problems triggered during the Anthropocene from a scientific perspective of peace ecology. These prolegomena need both thorough conceptual theoretical reflections and empirical research in the years to come, from both the peace and the environmental research communities as part of a combined effort across disciplines.

1.2 Addressing Global Environmental Challenges from a Peace Ecology Perspective

Already in 1896, the Swedish physicist Swante Arrhenius (1859–1925) hypothesized a link between the industrial burning of coal and an increase in atmospheric carbon dioxide (CO_2) concentration. Yet it was over 75 years before physicists, chemists and meteorologists met in Austria to discuss issues of *global environmental change* (GEC) and *climate change* (CC). This started the gradual *scientization* of problems and issues of GEC and CC. The scientization turned into a politicization in 1988, when the US Reagan Administration put climate change on the agenda of the G-7 in Toronto. This resulted in the *United Nations Framework Convention on Climate Change* (UNFCCC) in 1992, and five years later in 1997 in the adoption of the *Kyoto Protocol* (KP) in Japan. Since the start of the 21st century, the GEC and CC issues were discussed as (inter)national and human security issues or being *securitized* (Wæver 1995, 1997; Brauch 2009).

The *scientization* of the GEC, of CC issues and policy problems, was further advanced in the framework of four major GEC research networks: the *World Climate Research Programme* (WCRP) created in 1980 (Church et al. 2011), the establishment of the *International Geosphere–Biosphere Programme* (IGBP) in 1986 (Noone et al. 2011); of *DIVERSITAS* since 1991 (Walther et al. 2011) focusing on biodiversity research; the *International Human Dimensions Programme* since 1997 (IHDP; von Falkenhayn et al. 2011; Arizpe et al. 2016; Ehlers 2016) that brought the social sciences (geographers, political scientists et al.) into global change research.

The *Earth Systems Science Programme* (ESSP) was set up in 2001, superseded by the new *Future Earth Programme* in 2013,[1] is a ten-year International Research Initiative for Global Sustainability that builds on these four GEC research programmes and the ESSP (Leemans et al. 2011).

Brauch and Oswald Spring (2009), Brauch (2009) have developed the so-called *PEISOR* model that helps analyse the linkages between *pressure* (of the human and earth systems), their *effects* (in terms of environmental scarcity, degradation and stress), its physical *impacts* (e.g. temperature rise, precipitation change, sea-level rise, extreme weather events), *societal outcomes* (crises, migration, conflicts) and political *response* (e.g. in the context of GEC diplomacy or national implementations strategies) and illustrated these linkages in a study for UNCCD (Brauch/Oswald Spring 2009) on soil security and on migration from Mexico to the USA (Oswald Spring 2012).

The problems, issues and concerns of GEC and CC were addressed from different research programmes in political science and international relations, i.e. from (a) environmental, (b) security, (c) development and—with some delay—also from (d) peace studies. The authors argue that from the perspective of a normative approach of 'peace ecology in the Anthropocene', the classic issues of GEC including biodiversity, water, soil and climate, and their consequences for water, food, energy and waste can be fruitfully examined.

[1]See: http://www.icsu.org/future-earth.

1.3 Organization of the Book: Biodiversity, Water, Food, Energy and Waste

Much research has been conducted within the natural sciences on *biodiversity* (biologists, zoologists, plant and landscape specialists), *water* (hydrologists, geologists, engineers), *food* (agricultural specialists), *energy* (engineers, economists, political scientists) and *waste* (engineers on toxic and radiation issues) and to a lesser extend from the *social and political sciences* (including from peace studies).

In this book six chapters address partly issues of the scientific context that has been outlined above and specific issue areas of their research. The authors have been trained in different disciplines and represent a variety of interests and foci of ecology and peace studies that does not allow for a systematic series of related research questions or contribution to a single research programme or project. Therefore, this selection is rather accidental and eclectic, and offers a small snapshot of themes that peace researchers who are interested in ecology and peace issues are studying. These common interest is what binds the individual chapters together. These six chapters by authors from Australia, Canada, Finland, Germany and Mexico are multidisciplinary and address multiple perspectives.

Hans Günter Brauch—a retired political scientist, historian and international lawyer from Germany—offers a typology of time starting with cosmic and geological time of the universe and earth system and four human induced times: technical, structural, cyclical time and the short duration of structure creating or changing events. On the background of IPRA's 50th anniversary meeting—100 years after 1914 and 25 years after the collapse of the Soviet Union—the chapter discusses specific turning points during the short and turbulent 20th century between 1914 and 1989/1990.

The second chapter is by *Juliet Bennett,* a Ph.D. Candidate from the University of Sydney. Bennett discusses the "Global Ecological Crisis" as a form of structural violence, applying a conceptual linkage between Johan Galtung's "structural theory of imperialism" and Alfred Kahn's "tyranny of small decisions". This chapter closes by taking an imaginative leap, envisioning an idyllic alternative structure of a more peaceful and ecological world system.

This is followed by a chapter by *Katharina Bitzker*—a medical doctor and psychologist from Germany and Ph.D. candidate in Peace Studies at the University of Manitoba in Canada. Bitzker's chapter considers ways of "Loving Nature", discussing "The Emotional Dimensions of Ecological Peacebuilding". The fourth chapter is written by *Henri Myrttinen*—a political scientist, who was born in Finland, received some of his academic training in South Africa and works with International Alert in London. Myrttinen analyses "Drowning in complexity? Preliminary findings on addressing gender, peacebuilding and climate change in Honduras".

The chapter by *Úrsula Oswald Spring*—an ecologist from Mexico, with a PhD in anthropology and ecology from Zürich University—offers a critical review of the policy and scientific nexus debate on: "The Water, Energy, Food and Biodiversity

Nexus: New Security Issues in the Case of Mexico". A final chapter by *Brauch* discusses whether strategies of sustainability transition may enhance the prospects for achieving the goal of a sustainable peace in the Anthropcone.

In Chap. 2, *Hans Günter Brauch* discusses six historical times. *Cosmic* and *geological* time are concepts used in the history of the universe and earth. The *technical* (in the framework of 'technical revolutions') and *structural* times (in the context of 'international orders') can hardly be modified by governments and policymakers. The *conjunctural* time (in economics and politics) and shortlived *events* have in some cases become triggers of turning points. The assassination of Archduke Franz Ferdinand of Austria on 28 June 1914 in Sarajevo, the German attack on Poland on 1 September 1939 and the accidental opening of the Berlin Wall on 9 November 1989 fundamentally changed global *structures* that are usually beyond the influence of policymakers in office.

These six historical times and changing global contexts, political turning points, global transformations and transitions are discussed for international orders in the 20th century. It argues that the industrial revolution triggered the silent transition in geological time that resulted in the global transformation of the technological, economic and political systems and of international relations. The centennial catastrophe of 1914 led to World War I and the order of Versailles collapsed with the outbreak of World War II. The global peaceful change of 1989 did not result in a period of sustainable peace but in a new global disorder and global environmental challenges.

This chapter tries to introduce a *typology of time* as used in modern physics, in geology and earth sciences, archaeology and history of sciences and by the French school of structural history. It was triggered by the revolutionary observation of the Dutch atmospheric chemist and Nobel Laureate, Crutzen (2002) that "we are in the Anthropocene" or that 'we' as a member of the human species have for the first time directly interfered into the earth system and are in the process of modifying the conditions of our existence as 'humankind'.

This chapter argues that the human-induced silent transition in geologic time from the Holocene to the Anthropocene requires a fundamental rethinking of 'time' in the natural and social sciences and of 'peace in the Anthropocene'. This silent transition is not yet reflected in mainstream thinking in the social sciences, international relations, peace studies and in peace ecology. The discussion of turning points in the short twentieth century concluded that with the end of the Cold War the causes and the impacts of global environmental change and climate change have for the first time been put on the policy agenda.

Despite the setback in Copenhagen at COP 15 of the UNFCCC in December 2009 and the slowly moving and partly paralyzed global climate diplomacy, the outcome of COP 21 in Paris has given global climate diplomacy a new push. While this is necessary, it will not be sufficient. What is needed is a new "scientific revolution towards sustainability" and a new scientific worldview as a driver for policies aiming at the implementation of goals aiming at sustainable development with strategies aiming at a sustainability transition possibly contributing to a realization of a 'sustainable peace' in the Anthropocene.

In Chap. 3, *Juliet Bennett* has conceptually combined Galtung's "Structural Theory of Imperialism" (Galtung 1971) with Alfred Kahn's "Tyranny of Small Decisions" (Kahn 1966) arguing that the synthesis of theories sheds light on the multi-levelled and multi-directional influence of individuals, nations, institutions and culture.

Countless 'small decisions', that appear separate and distant from their collective long-term global consequences, are posited to be a root cause of the crisis. Solving the crisis calls for a holistic re-orienting of decision-making by people across many sectors of society aimed at long-term global interests rather than short-term personal interests. Examples of these decisions are considered. The chapter closes by imagining what a just and sustainable world system operating within planetary boundaries might look like, and consider examples of the type of decision-making it might involve.

In Chap. 4, *Katharina Bitzker* argues that while the mainstream environmental discourse seems to have taken a technocratic turn during recent years and promotes shallow ecological solutions which often fail to address underlying emotions as drivers for structural and cultural violence, there is also plenty of evidence of emerging research that offers a broader, more holistic perspective and puts the emotional/affective component—some call it love of nature—at its centre.

In this essay Bitzker explores how the experience of 'loving nature' has been conceptualized in some of the literature pertaining to cultural ecology so far and how these experiences translate (or do not translate) into different daily practices that are conducive to ecological peacebuilding and ultimately a 'happy planet'. Drawing on the work of anthropologist Kay Milton, one of the core questions becomes: is it a mere coincidence who is actively engaged and concerned with the well-being of nature and who might be more or less indifferent to the current ecological degradation? Loving (or at least respecting) nature and acting accordingly appears to be a prerequisite for love between humans at this point in time. The current global ecological degradation reminds us that focusing on human-human aspects of love alone tends to neglect the simple fact that we are destroying what gives us life—while being proud of our loving behaviour towards other human beings.

In this essay Bitzker highlights why it might be important to broaden current anthropo-centric models of love and shift to an ecological model of loving, how practices of resistance and complicity are embedded in an emotional field, why some sort of value coordinate system for 'sustainable loves' might be needed in the global north and the importance of embodiment or embodied emotions for our capacity to experience love or feel cut off from love.

In Chap. 5, *Henri Myrttinen* argues that although the interconnected issues of gender, climate change and peacebuilding have been high on the international agenda for the past decades, and the interplay between the issues has been researched in pairs (i.e. gender/peacebuilding, gender/climate change, climate change/peacebuilding), there is little research available on the simultaneous interplay of the three.

This chapter examines first some of the theoretical issues and dynamics at play when addressing these issues, followed by an overview of preliminary findings from Honduras. These findings are based on scoping research carried out by and for

International Alert, in preparation for a project on the gendered impacts of climate change-induced coffee rust and peacebuilding approaches to help communities cope with these.

In Chap. 6, *Úrsula Oswald Spring* analyses the security nexus between *water, energy, food and biodiversity* (WEF&B). The research question is, how could the nexus between WEF&B security be improved in a country with high environmental and social vulnerability, and which is seriously affected by climate change and organized crime? After a short conceptual review of WEF&B security, the dominant nexus is explored for Mexico, addressing first the feedbacks between water and biodiversity, and later changes in land use, food production and social vulnerability.

Mexico is an oil-exporting country and has the fourth most important reserve of shale gas in the world. It has extensive drylands where seventy-seven per cent of the population lives. These produce eighty-seven per cent of the GDP but receive only thirty-one per cent of the water that falls as rainwater; the environment and the aquifers are thus overexploited. Furthermore, a neo-liberal free trade policy has allowed highly subsidized food imports, as well as rural–urban and international migration of peasants. Finally, extreme events influenced by climate change, such as hurricanes and droughts, have had a negative impact on human lives and on the economy. In addition, organized crime controls most of the trade in migrants, drugs, and arms, as well as timber. A weak legal system has fostered small-scale crime, and this has increased public insecurity. As well as this, fracking activities in water-scarce regions are impacting on deep aquifers and limiting processes of adaptation to climate change in desert regions. The nexus between scarce water, overexploited aquifers, deforested areas, disasters, high food prices, weak rural government support, high energy prices and fragile governance is increasing poverty and the migration of farmers on rainfed lands, as well as creating the risk of social instability in urban areas.

Chapter 7 by *Hans Günter Brauch* focuses on hypothetical implications of uncertain outcomes of a long-term transformative change to achieve a sustainable development through a process of a sustainability transition. It addresses the question whether a long-term transformative change may result in a more peaceful environment.

The chapter is structured in ten parts. After a brief introduction, it discusses sustainable development as a goal and sustainability transition as a transformative process. It reviews the scientific debate on sustainability transition and its impact on a report on *A Social Contract for Sustainability*, examines the climate and energy policy initiatives of the European Union and analyses policy debates on climate and energy policy issues. The argument takes up the consequences of the human intervention into the Earth system that we are threatening the survival of humankind. The sustainable 'peace concept' is briefly conceptualized for the Anthropocene, whose realization requires major innovations in economic and environment policy. It points to contested visions, strategies and policies aiming at a sustainable peace with the goal to avoid security implications of climate change and counter resource conflicts and it concludes with a discussion of the need to develop strategies and policies of sustainability transition for a 'sustainable peace' in the Anthropocene.

This chapter argues why a long-term transformative change toward sustainability in the framework of a low carbon economy and society may result in a more peaceful environment, while all previous long-term changes in human history resulted in more deadly forms of warfare. As human beings have directly interfered into the Earth system since the industrial revolution they have become both the 'cause' but also the 'victims' of the consequences of global environmental and climate change, and they can also become part of the solution. This requires major changes in our own values, preferences and consumptive behaviour based on alternative pathways to achieve sustainable sectoral policies during this century to realize the goals of low carbon economy.

Among social scientists there is a need to overcome the professionalization through over-specialization and to enter into a dialogue between environmental studies and peace research. The scientific debate on 'sustainability transition' addresses multiple scientific, societal, economic, political and cultural needs to reduce GHG emissions not only by legally binding quantitative emission reduction obligations but also by unilateral bottom-up initiatives.

The suggested concept of 'peace ecology in the Anthropocene' still needs much theoretical reflection to transform it gradually from a conceptual idea into a possible research paradigm what would require a closer cooperation of both environmental and peace scholars.

In the years to come, 'peace ecology' may become a theme of specific degree programmes at universities around the globe, such as 'geo-ecology' had become in geography. Such efforts require multi-, inter- and if possible even transdisciplinary approaches (Oswald Spring et al. 2008) from the natural and social sciences with new scientific concepts, approaches, models, and theories that cross the boundaries between the narrow disciplinary analyses and assessments that still prevail in the organization and funding of scientific research.

The American biologist Wilson (1998) noted a growing *consilience* (interlocking of causal explanations across disciplines) in which the "interfaces between disciplines become as important as the disciplines themselves" that would "touch the borders of the social sciences and humanities." If 'peace ecology in the Anthropocene' should emerge from a conceptual idea to a new research paradigm it may contribute to the growing *consilience* Wilson observed.

References

Amster, Randall, 2015: *Peace Ecology* (Boulder, CO: Paradigm).

Arizpe, Lourdes; Price, Martin F.; Worcester, Robert, 2016: "The First Decade of Initiatives for Research on the Human Dimensions of Global (Environmental) Change (1986–1995)", in: Brauch, Hans Günter; Oswald Spring, Úrsula; Grin, John; Scheffran, Jürgen (Eds.): *Sustainability Transition and Sustainable Peace Handbook.* Hexagon Series on Human and Environmental Security and Peace 10 (Cham–Heidelberg–New York–Dordrecht–London: Springer International Publishing).

Boulding, Kenneth E., 1966: "The Economics of the Coming Spaceship Earth", in: Jarrett, H. (Ed.): *Environmental Quality in a Growing Economy* (Baltimore: Johns Hopkins Press).

Brauch, Hans Günter, 2009: "Securitzing Global Environmental Change", in: Brauch, Hans Günter; Oswald Spring, Úrsula; Grin, John; Mesjasz, Czeslaw; Kameri-Mbote, Patricia; Behera, Navnita Chadha; Chourou, Béchir; Krummenacher, Heinz (Eds.): *Facing Global Environmental Change: Environmental, Human, Energy, Food, Health and Water Security Concepts* (Berlin–Heidelberg–New York: Springer): 65–102.

Brauch, Hans Günter; Oswald Spring, Úrsula, 2009: *Securitizing the Ground—Grounding Security* (Bonn: UNCCD, 2009); at: http://www.unccd.int/Lists/SiteDocumentLibrary/Publications/dldd_eng.pdf.

Church, John A.; Asrar, Ghassem R.; Busalacchi, Antonio J.; Arndt, Carolin E., 2011: "Climate Information for Coping with Environmental Change: Contributions of the World Climate Research Programme", in: Brauch, Hans Günter; Oswald Spring, Úrsula; Mesjasz, Czeslaw; Grin, John; Kameri-Mbote, Patricia; Chourou, Béchir; Dunay, Pal; Birkmann, Jörn (Eds.), 2010: *Coping with Global Environmental Change, Disasters and Security—Threats, Challenges, Vulnerabilities and Risks* (Berlin–Heidelberg–New York: Springer-Verlag): 1257–1270.

Crutzen, Paul J., 2002: "Geology of Mankind", in: *Nature*, 415,3 (January): 23.

Crutzen, Paul J., 2011: "The Anthropocene: A geology of mankind", in: Brauch, Hans Günter; Oswald Spring, Úrsula; Mesjasz, Czeslaw; Grin, John; Kameri-Mbote, Patricia; Chourou, Béchir; Dunay, Pal; Birkmann, Jörn (Eds.): *Coping with Global Environmental Change, Disasters and Security—Threats, Challenges, Vulnerabilities and Risks* (Berlin–Heidelberg–New York: Springer-Verlag): 3–4.

Crutzen, Paul J.; Stoermer, Eugene F., 2000: "The Anthropocene", in: *IGBP Newsletter*, 41: 17–18.

Dalby, Simon, 2013a: "The Geopolitics of Climate Change", in: *Political Geography*, 37: 38–47.

Dalby, Simon, 2013b: "Climate Change: New Dimensions of Environmental Security", in: *RUSI Journal*, 158,3 (June/July): 34–43.

Dalby, Simon, 2014: "Rethinking Geopolitics: Climate Security in the Anthropocene", in: *Global Policy*, 5,1: 1–9.

Dalby, Simon, 2015: "Climate Geopolitics: Securing the Global Economy", in: *International Politics*, 52,4: 426–444.

Ehlers, Eckart, 2016: "From HDP to IHDP: Evolution of the International Human Dimensions of Global Environmental Change Programme (1996–2014)", in: Brauch, Hans Günter; Oswald Spring, Úrsula; Grin, John; Scheffran; Jürgen (Eds.): *Handbook on Sustainability Transition and Sustainable Peace* (Cham–Heidelberg–New York–Dordrecht–London: Springer International Publishing).

Galtung, Johan, 1971: "A Structural Theory of Imperialism", in: *Journal of Peace Research*, 8,2: 81–117.

Kahn, Alfred E., 1966: "The Tyranny of Small Decisions: Market Failures, Imperfections, and the Limits of Economics", in: *Kyklos*, 19,1: 23–47.

Kyrou, Christos N., 2007a: "Peace Ecology: An Emerging Paradigm in Peace Studies", in: *The International Journal of Peace Studies*, 12,2 (Spring/Summer): 73–92.

Kyrou, Christos N., 2007b: "Methodological Perspectives from Peace Ecology: An Emerging Paradigm in Peace Studies", Conference on Cutting Edge Theories and Recent developments in Conflict Resolution, Syracuse, Syracuse University, The Maxwell School, Program on the Analysis and Resolution of Conflicts, 27–28 September.

Leemans, Rik; Rice, Martin; Henderson-Sellers, Ann; Noone, Kevin, 2011: "Research Agenda and Policy Input of the Earth System Science Partnership for Coping with Global Environmental Change", in: Brauch, Hans Günter; Oswald Spring, Úrsula; Mesjasz, Czeslaw; Grin, John; Kameri-Mbote, Patricia; Chourou, Béchir; Dunay, Pal; Birkmann, Jörn, 2010: *Coping with*

Global Environmental Change, Disasters and Security—Threats, Challenges, Vulnerabilities and Risks (Berlin–Heidelberg–New York: Springer-Verlag): 1205–1220.

Noone, Kevin J.; Nobre, Carlos; Seitzinger, Sybil, 2011: "The International Geosphere–Biosphere Programme's (IGBP) Scientific Research Agenda for Coping with Global Environmental Change", in: Brauch, Hans Günter; Oswald Spring, Úrsula; Mesjasz, Czeslaw; Grin, John; Kameri-Mbote, Patricia; Chourou, Béchir; Dunay, Pal; Birkmann, Jörn (Eds.), 2010: *Coping with Global Environmental Change, Disasters and Security—Threats, Challenges, Vulnerabilities and Risks* (Berlin–Heidelberg–New York: Springer-Verlag): 1249–1256.

Oswald Spring, Úrsula, 2012: "Environmentally-Forced Migration in Rural Areas: Security Risks and Threats in Mexico", in: Scheffran, Jürgen; Brzoska, Michael; Brauch, Hans Günter; Link, Peter Michael; Schilling, Janpeter (Eds.): *Climate Change, Human Security and Violent Conflict: Challenges for Societal Stability* (Berlin–Heidelberg–New York: Springer): 315–350.

Oswald Spring, Úrsula; Brauch, Hans Günter, 2008: "Reconceptualizing Security in the 21st Century: Conclusions for Research and Peacemaking", in: Brauch, Hans Günter; Oswald Spring, Úrsula; Mesjasz, Czeslaw; Grin, John; Dunay, Pal; Behera, Navnita Chadha; Chourou, Béchir; Kameri-Mbote, Patricia; Liotta, P.H. (Eds.): *Globalization and Environmental Challenges: Reconceptualizing Security in the 21st Century* (Berlin–Heidelberg–New York: Springer-Verlag): 941–954.

Oswald Spring, Úrsula; Brauch, Hans Günter; Tidball, Keith G. (Eds.), 2014a: *Expanding Peace Ecology: Peace, Security, Sustainability, Equity and Gender – Perspectives of IPRA's Ecology and Peace Commission* (Cham–Heidelberg–New York–Dordrecht–London: Springer).

Oswald Spring, Úrsula; Brauch, Hans Günter; Tidball, Keith G., 2014b: "Expanding Peace Ecology: Peace, Security, Sustainability, Equity and Gender", in: Oswald Spring, Úrsula; Brauch, Hans Günter; Tidball, Keith G. (Eds.): *Expanding Peace Ecology: Peace, Security, Sustainability, Equity and Gender—Perspectives of IPRA's Ecology and Peace Commission* (Cham–Heidelberg–New York–Dordrecht–London: Springer): 1–32.

von Falkenhayn, Louise; Rechkemmer, Andreas; Young, Oran R., 2011: "The International Human Dimensions Programme on Global Environmental Change—Taking Stock and Moving Forward", in: Brauch, Hans Günter; Oswald Spring, Úrsula; Mesjasz, Czeslaw; Grin, John; Kameri-Mbote, Patricia; Chourou, Béchir; Dunay, Pal; Birkmann, Jörn, 2010: *Coping with Global Environmental Change, Disasters and Security—Threats, Challenges, Vulnerabilities and Risks* (Berlin–Heidelberg–New York: Springer-Verlag): 1221–1234.

Wæver, Ole, 1995: "Securitization and Desecuritization", in: Lipschutz, Ronnie D. (Ed.): *On Security* (New York: Columbia University Press): 46–86.

Wæver, Ole, 1997: *Concepts of Security* (Copenhagen: Department of Political Science).

Walther, Bruno A.; Larigauderie, Anne; Loreau, Michel, 2011: "DIVERSITAS: Biodiversity Science Integrating Research and Policy for Human Well-Being", in: Brauch, Hans Günter; Oswald Spring, Úrsula; Mesjasz, Czeslaw; Grin, John; Kameri-Mbote, Patricia; Chourou, Béchir; Dunay, Pal; Birkmann, Jörn (Eds.), 2010: *Coping with Global Environmental Change, Disasters and Security—Threats, Challenges, Vulnerabilities and Risks* (Berlin–Heidelberg–New York: Springer-Verlag): 1235–1248.

Wilson, Edward O., 1998: *Consilience* (New York: Knopf).

Chapter 2
Historical Times and Turning Points in a Turbulent Century: 1914, 1945, 1989 and 2014?

Hans Günter Brauch

Abstract This chapter discusses six historical times. *Cosmic* and *geological* time are concepts used in the context of the history of the universe and earth. *Technical* and *structural* times (in the framework of 'technical revolutions' and 'international orders') can hardly be modified by governments and policymakers. *Conjunctural* time (in economics and politics) and short-lived *events* have in some cases become triggers for turning points. They fundamentally change global *structures* that are usually beyond the influence of policymakers in office. These six historical times and changing global contexts, political turning points, global transformations and transitions are discussed for international orders in the twentieth century. The chapter argues that the Industrial Revolution triggered the silent transition in geological time that resulted in the global transformation of technological, economic and political systems and of international relations. The catastrophe of 1914 led to World War I and the order of Versailles collapsed with the outbreak of World War II. The global peaceful change of 1989 resulted not in a period of sustainable peace but in a new global disorder and in global environmental challenges. It is uncertain whether humankind will understand the consequences of a situation where "we are in the Anthropocene" and "we are the threat" and "we alone can become the solution".

Keywords Cosmic time · Geological time · Technical time · Structural time · Conjunctural time · Events · Turning point · Neolithic revolution · Industrial revolution · Holocene · Anthropocene · Global transformation

---

I am grateful for critical comments and valuable suggestions to Ms Juliet Bennett, University of Sydney, Australia; Dr. Carl Bruch, Washington DC, USA; Prof. Dr. Simon Dalby, Waterloo, Canada; Prof. Dr. Kalevi Holsti, Vancouver, Canada; Prof. Dr. Ken Conca, Washington DC, USA. I am also grateful to Nobel laureate Prof. Dr. Paul J. Crutzen (2016), who taught me more about the human predicament with his work on the nuclear winter, the Anthropocene and the sustainability revolution than my own discipline.

PD Dr. Hans Günter Brauch, chairman, Peace Research and European Security Studies (AFES-PRESS) from 1987; co-convenor, IPRA's Ecology and Peace Commission (2012–2016), Mosbach, Germany; email: brauch@afes-press.de.

H.G. Brauch et al. (eds.), *Addressing Global Environmental Challenges from a Peace Ecology Perspective*, The Anthropocene: Politik—Economics—Society—Science 4, DOI 10.1007/978-3-319-30990-3_2

2.1 Introduction

The author's conclusion is thus that "we are the threat" to our own survival as a species but also that only we as human beings can be and become the solution. This is provided that humankind becomes fully aware of the revolutionary consequences of the silent transition, a transition that has taken place since the 'Industrial Revolution' and that has accelerated during the seventy years since the end of World War II. The powerful economic and political interests and ideologies supported by adherents of *business as usual*, and trivialization by elements of the mass media—all these have ignored, ridiculed or attacked and attempted to counter these revolutionary insights and their conceptual consequences. The many turning points in the political history of events during the short twentieth century and during the young twenty-first century, as we see daily in the news, conceal the fact that humankind may, during the twenty-first century, face for the first time a global "survival dilemma" (Brauch 2008). A conceptual debate is needed among social scientists, especially from those who specialize in ecology and peace issues, as we seem to be slowly and silently undermining the very conditions of our human existence.

In this much broader scientific context, this chapter focuses on 'time' in *cosmic*, *geological* and *human history* and on specific 'turning points' in political history since the end of the 'long nineteenth century' (1789–1914) and during the 'short twentieth century' (1914–1989/1991).[1] The latter period started with the outbreak of World War I in July 1914, the catastrophe. A major turning point was the German attack on Poland on 1 September 1939 that triggered World War II, a war which resulted in global political and economic transformation since its end in 1945. The century symbolically ended with the fall of the Berlin Wall in 1989 and the implosion of the Soviet Union in 1991 that terminated the cold war (1947–1989/1991).

A century after the outbreak of World War I and about twenty-five years after the end of the cold war, humankind has seen no 'peace dividend', and no lasting or 'sustainable' peace has emerged. Rather, 'new wars' (Kaldor 1997; Münkler 2004) have replaced the 'proxy wars' of the superpowers in the global South. Numerous new insecurities have been observed, and a reconceptualization of security has occurred (Brauch et al. 2008, 2011; Brauch et al. 2009). Peace research has thus not become obsolete. However, many new global political, economic, and environmental challenges require a conceptual rethinking of 'peace and security' as well as

---

[1]Inspired by Fernand Braudel's idea of the 'long sixteenth century' (*c.*1450–1640), the concept of the 'long nineteenth century', was according to Eric Hobsbawm the period between the years 1789 and 1914 that he had analysed in *The Age of Revolution: Europe 1789–1848* (1962), *The Age of Capital: 1848–1875* (1975), and *The Age of Empire: 1875–1914* (1987). In *The Age of Extremes: The Short Twentieth Century, 1914–1991* (1994), Hobsbawm discussed the short twentieth century (1914–1991). See for the German historical debate: Bauer (2004), Kocka ([10]2002), Osterhammel (2009).

of 'ecology' and 'peace' in a new era of earth and human history, the Anthropocene.

The thesis of this chapter—inspired by Crutzen (2002, 2011)—is that humankind has become aware of a far more fundamental turning point in earth history, with the transition from the geological period of the 'Holocene' to the 'Anthropocene'. This transition is an outcome of the first major human intervention in the earth system, as a result of human consumption of fossil energy sources since the Industrial Revolution and most particularly during the past seventy years, through the burning of coal, oil and gas. Humankind is starting to understand that "we are in the Anthropocene!" and that "we are the threat!" caused by the societal outcomes of the physical impacts of anthropogenic global environmental and climate change. But "we can and must also be the solution", in order to avoid the long-term negative and potentially violent consequences of global environmental and climate change by the end of this century.

The social construction of this new global reality has gradually been 'scientized' since the 1970s, 'politicized' since 1988, and 'securitized' since the early twenty-first century (Brauch 2002, 2009). The implication of this 'silent transition' may require a new 'scientific' or 'Copernican revolution' (Kuhn 1962), with a new 'world view of sustainability' (Clark et al. 2004) or a 'new social contract for sustainability' (WBGU 2011).

The discussion of the impact on issues of security of this silent transition to the Anthropocene era of earth history (Dalby 2013a, b, 2014, 2015) and on 'sustainable peace' (Brauch 2016; Chap. 7 in this book) is gradually emerging in the social sciences in general and in ecology and peace studies in particular. But the initial conceptualizations of 'peace ecology' (Kyrou 2007; Amster 2014) do not yet reflect the need for the framing of a new 'peace ecology in the Anthropocene' (Oswald Spring et al. 2014).

The human-induced silent transition in geological time from the Holocene to the Anthropocene also requires a fundamental rethinking of the 'role of time' in the natural and social sciences and in the humanities, including peace research (Reychler 2015a, b), as well as in the political, social and economic systems and on the part of its key actors. This is addressed in the *Handbook on Sustainability Transition and Sustainable Peace* and discussed below in "Building Sustainable Peace by Moving towards Sustainability Transition".

This chapter is structured in ten parts. After this introduction six historical times and changing global contexts are introduced (2.2), then the chapter discusses political turning points, global transformations and transitions for international orders in the twentieth century (2.3), examines the Industrial Revolution as a trigger for the silent transition in geological time (2.4) that resulted in the global transformation of technological, economic and political systems and of international relations (2.5), where the catastrophe of 1914 led to World War I (2.6) and the order of Versailles collapsed with the outbreak of World War II (2.7). The uniquely peaceful global change of 1989 did not, however, result in a period of sustainable peace but in a new global disorder with new asymmetric wars and new global environmental challenges (2.8). A century after 1914, the present is not a turning point and it is still uncertain whether humankind, as represented by its governments and policymakers, will understand the consequences of a situation where "we are in the Anthropocene" (2.9) and that "we are now the threat" and that "only we can become the solution" (2.10).

2.2 Historical Times and Changing Global Contexts

In the natural and social sciences six historical times with different turning points can be distinguished that have triggered numerous contextual changes. However, it is unlikely that a century after the catastrophe of 1914 a similar turning point can be observed in 2014 or 2015 (as was the case in 1914, 1945 or 1989), nor whether the new conflict between Russia and the West over the annexation of the Crimea and Russia's role in the Ukraine or the role of the new *Islamic Caliphate* (IS) in parts of Syria and Iraq will fundamentally change international relations.

The American–Chinese climate initiative of 2014 and the change to more sustainable energy policies announced by Presidents Hollande (July 2014) and Obama (August 2015) have contributed to overcoming the paralysis of global climate diplomacy that has lasted from the *Conference of Parties* (COP: COP15) of the *United Nations Framework Convention on Climate Change* (UNFCCC) in 2009 in Copenhagen to COP21 in Paris in December 2015. If the Paris Agreement (2015) should fail to be implemented nationally as no legally binding climate agreement has evolved, humankind may well face a 4 °C world (Schellnhuber et al. 2012, 2016) with 'catastrophic climate change' (Schellnhuber et al. 2006), and this would significantly raise the economic costs of non-action (Stern 2006, 2009).

2.2.1 The Term and Concept of 'Time'

The 'term' and 'concept' of time (*chronos* vs. *kairos* [Greek], *tempus* [Latin], *temps* [French], *Zeit* [German]) has been widely used with a range of meanings in everyday speech and in scientific analyses in different disciplines (natural sciences [relativity, cosmology, physics, biology], geology [stratigraphy], humanities [philosophy, history] and in the social sciences [political science, economics]).

The *Shorter Oxford English Dictionary* ([5]2002, volume 2: 3272–3273) defines 'time' as a "finite extent of continued existence; e.g. the interval between two events, or the period during which an action or state continues" … "a period in history, a period in the existence of the world; an age, an era"; as "a point in time; a space of time treated without ref. to duration"; and in a generalized sense as "duration conceived as having a beginning and an end; finite duration distinct from eternity".

The *Encyclopaedia Britannica* ([15]1998, volume 28: 662–673) notes that "one facet of human consciousness is the awareness of time. … People feel, think, and act in the time flow". The essay discusses the role of "time and its history of thought and action" … "the contemporary philosophies of time", and "time as systematized in modern scientific society". The latest edition of the German *Brockhaus Enzyklopädie* ([21]2006, volume 30: 486–495) distinguishes in its survey between philosophical considerations of the nature of time, time in classical mechanics and in the era of the theory of relativity, the start and end of time, time arrows, time in religion and in consciousness, and social time.

The 'historical dictionary of philosophy' (*Historisches Wörterbuch der Philosophie*, volume 12: 1185–1262) reviews the development of the concept of time from the pre-Greek period, through the Greek–Roman era, the Middle Ages, Humanism and the Renaissance to Kant and Heidegger, as well as the use of time in physics, society and culture and concepts of time in cultures outside Europe (India, China, Japan).

In this chapter different concepts of time are used, according to their duration:

1. *Cosmic time*, used in models of *physical cosmology* (since Max Planck), and referring to time since the Big Bang about 13.8 billion years ago; this is beyond the scope of any collective human impact.
2. *Geological time*, which describes the timing of and relationships between events throughout the earth's history of about 4.54 billion years; its scales are adopted by geologists and earth scientists and defined by members of the *International Commission on Stratigraphy*. Its most recent accepted epoch is the *Holocene*, the period since the retreat of the Quaternary Ice Age some 12,000 years ago that made the rise of human civilizations possible.
3. The time of the *technical revolutions* (the 'Neolithic' or 'agricultural' revolution of 10,000–6,000 BCE, and the 'Industrial Revolution' from about 1750 CE and its different phases of innovation).

In claiming that "we are in the Anthropocene", Crutzen stated that since the Industrial Revolution humankind has for the first time directly interfered in the earth system, triggering complex processes of global environmental (soil, water, biodiversity) and climate change.

In human history the French social historian Fernand Braudel in his masterpiece *The Mediterranean and the Mediterranean World in the Age of Philip II* (1946, 1969, 1972) distinguished between three historical times: (a) long duration (*la longue durée*), (b) repeating historical cycles (*histoire de conjuncture*), and (c) events (*l'histoire événementielle*). Braudel's periodization is extensively used in history and in the social sciences (political science and economics). Other periodizations are common in economic history and theory (e.g. mercantilism, capitalism, socialism, neo-liberalism) and in feminist discourse (e.g. emergence of patriarchy, see: Oswald Spring 2016).

2.2.2 Cosmic Time: Beyond Human Intervention

According to the *Encyclopaedia Britannica* ([15]1998, volume 28: 663–664) a cyclic view of *cosmic* and human history prevailed among the Hindus, pre-Christian Greeks and Aztecs.

The Chinese, Hindus, and Greeks saw cosmic time as moving in an alternating rhythm, classically expressed in the Chinese concept of the alternating between Yin, the passive female principle, and Yang, the dynamic male principle … In the philosophy of Empedocles … the equivalents of Yin and Yang are Love and Strife. … Plato and Aristotle took it for granted that human society, as well as the cosmos, have been, and will continue to be, wrecked and rehabilitated any number of times.

> In modern physics *cosmic time* has been identified with the time since the Big Bang; it is commonly used in the Big Bang models of physical cosmology. It is defined for homogeneous, expanding universes as follows: Choose a time coordinate so that the universe has the same density everywhere at each moment in time (the fact that this is possible means that the universe is, by definition, homogeneous).[2]

Recently *physical cosmology* has been described as:

> the study of the largest-scale structures and dynamics of the Universe and [is] concerned with fundamental questions about its origin, structure, evolution, and ultimate fate. … Physical cosmology, as it is now understood, began with … Albert Einstein's general theory of relativity, followed by major observational discoveries in the 1920s: [by] Edwin Hubble … work by Vesto Slipher and others showed that the universe is expanding. These advances … allowed the establishment of the Big Bang Theory, by Georges Lemaitre, as the leading cosmological model. … Cosmology draws heavily on the work of many disparate areas of research in theoretical and applied physics. Areas relevant to cosmology include particle physics experiments and theory, theoretical and observational astrophysics, general relativity, quantum mechanics, and plasma physics.[3]

This *cosmic time* that refers to the 'Big Bang' of about 13.8 billion years ago is beyond any human impact and will thus not be considered further.

2.2.3 Geological Time: Transition from the Holocene to the Anthropocene

By claiming at a conference around the turn of the millennium in Cuernavaca (Mexico) that "We are in the Anthropocene!", the Nobel chemistry laureate Crutzen (2002, 2016) challenged the *geological time scale* (GTS) geologists and Earth

[2]See "Cosmic time"; at: https://en.wikipedia.org/wiki/Cosmic_time.

[3]See "Physical cosmology"; at: https://en.wikipedia.org/wiki/Physical_cosmology. See also: *Encyclopaedia Britannica* ([15]1998, vol. 28: 665–667); Balbi, Amedeo: "Cosmology and time"; at: http://arxiv.org/pdf/1304.3823v1.pdf; Rugh, S. E.; Zinkernagel, H.: "On the physical basis of cosmic time"; at: http://philsci-archive.pitt.edu/4020/1/CosmicTime.pdf (10 August 2015).

scientists have used to describe "the timing and relationships between events that have occurred throughout Earth's history" by relating "stratigraphy to time". According to the definition of the *International Commission on Stratigraphy*,[4] during the *Holocene* (from 11,700 BP to the present) these events have occurred:

> Quaternary Ice Age recedes, and the current interglacial begins; rise of human civilization. Sahara forms from savannah, and agriculture begins. Stone Age cultures give way to Bronze Age (3300 BC) and Iron Age (1200 BC), giving rise to many pre-historic cultures throughout the world.[5]

From a geological perspective, the *Holocene*

> encompasses the growth and impacts of the human species worldwide, including all its written history, development of major civilizations, and overall significant transition toward urban living in the present. Human impacts of the modern era on the Earth and its ecosystems may be considered of global significance for future evolution of living species, including approximately synchronous lithospheric evidence, or more recently atmospheric evidence of human impacts.[6]

The *International Commission on Stratigraphy* has not yet approved the concept of the 'Anthropocene', but in 2008 it set up a study group to examine it. For geologists the

> Anthropocene is an informal geologic chronological term for the proposed epoch that began when human activities had a significant global impact on the Earth's ecosystems. The term … was coined … in the 1980s by ecologist Eugene F. Stoermer and has been widely popularized by … Paul Crutzen, who regards the influence of human behaviour on the Earth's atmosphere in recent centuries as so significant as to constitute a new geological epoch for its lithosphere. … The Anthropocene has no precise start date, but based on atmospheric evidence may be considered to start with the Industrial Revolution.[7]

A proposal was presented to the Stratigraphy Commission of the Geological Society of London in 2008 "to make the Anthropocene a formal unit of geological epoch divisions", and the Stratigraphy Commission set up an independent working group of scientists from various geological societies "to determine whether the Anthropocene will be formally accepted into the Geological Time Scale" with the aim of reaching a decision by 2016.[8]

---

[4]See for background on the "geological time scale", at: http://en.wikipedia.org/wiki/Geologic_time_scale (21 January 2015).

[5]See at: http://en.wikipedia.org/wiki/Geologic_time_scale, based on these sources: "NASA Scientists React to 400 ppm Carbon Milestone" (15 January 2014).

[6]For the Holocene, see at: http://en.wikipedia.org/wiki/Holocene; Roberts (1998); Mackay et al. (2003).

[7]For the "Anthropocene", see at: http://en.wikipedia.org/wiki/Anthropocene; (Crutzen and Stoermer 2000; Crutzen 2002; Steffen et al. 2011; Zalasiewicz et al. 2008, 2010).

[8]See: "Anthropocene", at: http://en.wikipedia.org/wiki/Anthropocene; see the Working Group on the 'Anthropocene', International Commission on Stratigraphy (ICS); at: http://quaternary.stratigraphy.org/workinggroups/anthropocene/; see also the most recent "Newsletter of the Anthropocene Working Group", September 2014; at: http://quaternary.stratigraphy.org/workinggroups/anthropo/anthropoceneworkinggroupnewslettervol5.pdf.

Crutzen (2011: 3–4), in "The Anthropocene: a geology of mankind", summarized his observations on the human-induced interventions in and impacts on the earth system:

> Considering these and many other major and still growing impacts of human activities on earth and atmosphere, and at all scales, it thus is more than appropriate to emphasize the central role of humankind in the environment by using the term 'Anthropocene' for the current geological epoch.
>
> The impact of current human activities is projected to last and even expand over long periods.

In a more recent joint contribution Steffen et al. (2007: 614) argued:

> The term Anthropocene … suggests that the Earth has now left its natural geological epoch, the present interglacial state called the Holocene. Human activities have become so pervasive and profound that they rival the great forces of Nature and are pushing the Earth into planetary terra incognita. The Earth is rapidly moving into a less biologically diverse, less forested, much warmer, and probably wetter and stormier state.

Steffen et al. (2007: 616–618) distinguished two stages of the Anthropocene: (a) the *industrial era* (1800–1945), and (b) the *great acceleration* (1946–2015) (documented in Fig. 2.1):

> From 1950 to 2000 the percentage of the world's population living in urban areas grew from 30 to 50 % and continues to grow strongly. … The pressure on the global environment from this burgeoning human enterprise is intensifying sharply. Over the past 50 years, humans have changed the world's ecosystems more rapidly and extensively than in any other comparable period in human history … The Earth is in its sixth great extinction event, with rates of species loss growing rapidly for both terrestrial and marine ecosystems … The atmospheric concentrations of several important greenhouse gases have increased substantially, and the Earth is warming rapidly … More nitrogen is now converted from the atmosphere into reactive forms by fertilizer production and fossil fuel combustion than by all of the natural processes in terrestrial ecosystems put together. … The Great Acceleration took place in an intellectual, cultural, political, and legal context in which the growing impacts upon the Earth System counted for very little in the calculations and decisions made in the world's ministries, boardrooms, laboratories … The exponential character of the Great Acceleration is obvious from our quantification of the human imprint on the Earth System, using atmospheric CO_2 concentration as the indicator … Although by the Second World War the CO_2 concentration had clearly risen above the upper limit of the Holocene, its growth rate hit a take-off point around 1950. Nearly three-quarters of the anthropogenically driven rise in CO_2 concentration has occurred since 1950 (from about 310 to 380 ppm), and about half of the total rise (48 ppm) has occurred in just the last 30 years (Steffen et al. 2007: 616–618).

Steffen et al. (2007: 619–620) distinguished three philosophical responses: (a) *business as usual*, (b) *mitigation*, and (c) *geoengineering*, any of which "can raise serious ethical questions and intense debate", given "the possibility for unintended and unanticipated side effects that could have severe consequences. The

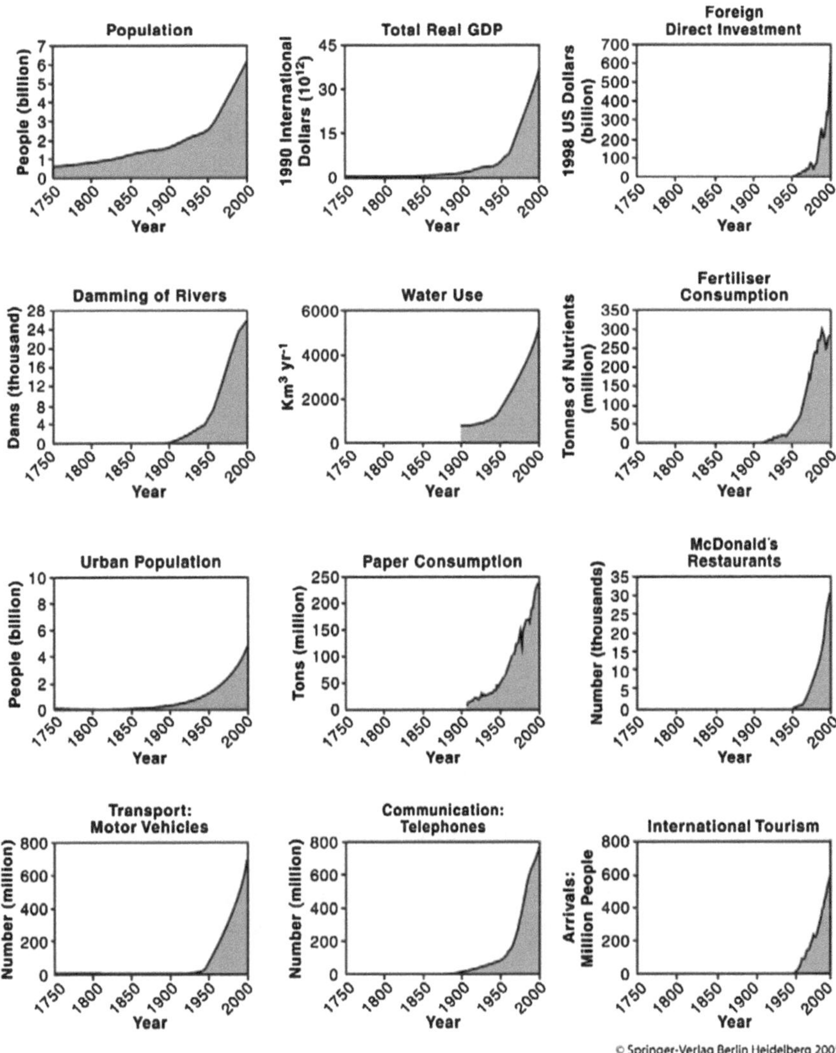

Fig. 2.1 The change in human enterprise from 1750 to 2000. The *great acceleration* is clearly shown in every component of human enterprise included in the figure. Either the component was not present before 1950 (e.g. foreign direct investment) or its rate of change increased sharply after 1950 (e.g. population). *Source* Steffen et al. (2007: 617), based on © Springer-Verlag (2005)

cure could be worse than the disease." They concluded that "The Great Acceleration is reaching criticality", but pointed to "evidence for radically different directions built around innovative, knowledge-based solutions".

2.2.4 Technical Time: Agricultural and Industrial Revolutions

The new concept of *technical time*, referring to the two major *technical revolutions*, has not yet been conceptualized by archaeologists, philosophers and historians of science and technical innovation. The concept of *technical time* refers to the numerous technical innovations and societal, economic and political changes brought about by the two major technical revolutions in human history so far, the *Neolithic* or *Agricultural Revolution* and the *Industrial Revolution* (1750–present), which have both caused 'great transformations' (Polanyi 1944).

The concept of the 'Neolithic Revolution' was coined by V. Gordon Childe in 1936 by analogy with 'Industrial Revolution', a term used by Friedrich Engels and L. A. Blanqui and given its present meaning by the British historian Arnold Toynbee, who used it to refer to the period of accelerated technological, economic and social change from 1769 onwards (Watt's invention of the steam engine) and to developments in Great Britain from 1785.

In the Holocene, the 'Neolithic' or 'first agricultural revolution' referred to the transition "from nomadic hunting and gathering communities and bands to agriculture and settlement"[9] that took place between twelve and six thousand years ago, first in the Fertile Crescent (Mesopotamia, Anatolia, Levant) and later in Europe. These changes meant the adoption of new farming techniques, crop cultivation, and the domestication of animals, and this fostered the development of permanent or semi-permanent settlements. The concept of land ownership emerged, as did a hierarchy in society; towns were founded and trading communities helped to feed them with the 'surplus production' from increasing crop yields while craftsmen such as potters supplied the needs of the evolving urban settlements and city-states and kingdoms.

According to Harlan (1992) the *Neolithic Revolution* took place for three reasons: domestication was brought about (a) for religious reasons; (b) because of crowding and stress; and (c) through discoveries made by the food-gatherers. Archaeologists distinguish three routes for the start of *Neolithization* in central Europe: (a) the classical route from the Near East via south-eastern Europe to central Europe, (b) from the steppes of Asia through the Russian forests to northern Europe with its ceramic traditions and (c) the western route across the Mediterranean to Europe (Gronenborn/Terberger 2014: 7; Sirocko [3]2012; Zimmermann 2009; Schmidt 2009).[10]

During the Neolithic age three agricultural crops developed as the basis for the diet of several advanced civilizations: cereals (wheat) in the *Fertile Crescent* from 9,000 BCE (Ethiopia, Egypt, Mesopotamia, Anatolia, Levant), rice in south-east Asia and in China from 8,000 BCE, and maize (corn) from 3,500 to 2,500 BCE in

[9]See "Neolithic Revolution"; at: https://simple.wikipedia.org/wiki/Neolithic_revolution (12 August 2015).

[10]Gronenborn/Terberger (2014), Sirocko ([3]2012), Zimmermann (2009), Schmidt (2009).

Meso-America (Mexico). With the division of labour (including that between men and women) a hierarchization of societies (including patriarchy) emerged, as well as writing and scripts, in Mesopotamia, China and Meso-America.

While human activities had an impact on the local environment, and the overuse of scarce water often contributed to the collapse of advanced civilizations (Egypt, Meso-America, China), these local impacts did not affect the earth system nor the global climate, as the greenhouse gas concentrations in the atmosphere remained stable throughout the Holocene era. This situation fundamentally changed due to the impact of the 'Industrial Revolution' during the 'industrial era' (1750–1945 or to the present) and the 'nuclear era' (1945–present).

The term 'Industrial Revolution' was coined by French diplomat Louis-Guillaume Otto (1799) and later used by Jérôme-Adolphe Blanqui in 1837 in his description of *la révolution industrielle*, while Friedrich Engels in *The Condition of the Working Class in England* (1844) referred to an Industrial Revolution which "changed the whole of civil society". The term was popularized by lectures by Arnold Toynbee (1881), while several historians stressed a gradual evolutionary change (Clapham 1926, 1936). The 'Industrial Revolution' points to a

> transition to new manufacturing processes … from … 1760 to … between 1820 and 1840. This transition included going from hand production methods to machines, new chemical manufacturing and iron production processes, improved efficiency of *water power*, the increasing use of *steam power*, and the development of *machine tools*. It also included the change from wood and other *bio-fuels* to *coal*. … The Industrial Revolution marks a major turning point in history … [It] began in Great Britain, and spread to Western Europe and North America within a few decades. [Its] precise start and end … is still debated among historians, as is the pace of economic and social changes.[11]

Many historians and social scientists have distinguished different phases of the Industrial Revolution caused by different inventions and waves of technological innovation: (a) the initial phase triggered by Watt's invention of the steam engine (1769) and its application in the textile industry (1760–1840) and (b) a second phase (1840–1870), "when technological and economic progress continued with the increasing adoption of steam transport (steam-powered railways, boats and ships), the large-scale manufacture of machine tools and the increasing use of machinery in steam-powered factories" (Ziegler 2009). Ziegler (2009) applied the term 'Industrial Revolution' to the period between 1760 and 1850, but he distinguished three phases of 'industrialization': (a) light industry, largely in the textile sector (1760–1840), (b) heavy industry, primarily railway construction (1830–1890), and (c) electrotechnical industries (since the 1890s).

The invention and application of electricity and of the telephone as well as the invention of the Otto motor and the car by Carl Benz (1886) and of new engines for ships and aircraft resulted in phases of energy, communication and transport

---

[11] See "Industrial Revolution", at: https://en.wikipedia.org/wiki/Industrial_Revolution.

revolutions. The process of technological innovation and its application accelerated during and in the immediate aftermath of World War II, with the USA as the model and pace-setter ('the American way of life'), until the European countries, Japan and (since Deng's reforms in the late 1970s) China became its competitors. As a consequence of these different phases of industrialization, a modern industrial society emerged during the twentieth century.

These changes in the economic system were accompanied by a 'scientific revolution', especially in the many disciplines of the basic natural sciences (physics, chemistry, biology) and in applied engineering (Jochum 2010); this triggered numerous changes in settlement patterns (the emergence of industrialized centres), in nutrition and diet, and in changing rhythms of life and work, and resulted in massive sociocultural processes of change. The social sciences and humanities focused increasingly on the related changes in societal relations, the value system and human behaviour in the industrial and nuclear era since 1945, once the atomic bomb and nuclear energy had added a new dimension to warfare and to the energy system.[12] This process increased the demand for imports of raw materials from all parts of the world and for the export of industrial mass production, and this intensified international trade and was an incentive for a revolution in the global system of transport.

The Industrial Revolution, the phases of industrialization and the waves of innovation (Fig. 2.2) were made possible by a shift from wood to fossil fuels (coal, oil and gas), and accompanied by an exponential increase in the consumption of fossil energy and (since the 1950s and 1960s) of nuclear energy as a source of electricity. The negative impacts on society have been gradually identified by natural scientists since the 1970s once they had started to analyse processes of global environmental change and climate change.

This debate has now resulted in a global consensus among the scientific community that 'anthropogenic' interventions in the earth system (IPCC 1990, 1995, 2001, 2007, 2014a) have brought about a fundamental change in earth history, and this is why Paul J. Crutzen has argued: "we are in the Anthropocene". Thus, for the first time in history, human activity, because of its impact during the Industrial Revolution (*technical time*), has triggered a change in *geological time*. As a result of an extended nuclear war, a 'nuclear winter' (Crutzen/Birks 1982) could also affect the atmosphere by darkening the sun with dust and thus cooling the earth, triggering bad harvests as happened after major eruptions of volcanoes throughout earth history. An example is the explosion of Krakatoa in Indonesia during the 1860s, which caused famine elsewhere.

---

[12]See the entries on: "Industrial Revolution" and "industrial era", in: *Brockhaus Enzyklopädie* ([21]2006, vol. 13: 260–261).

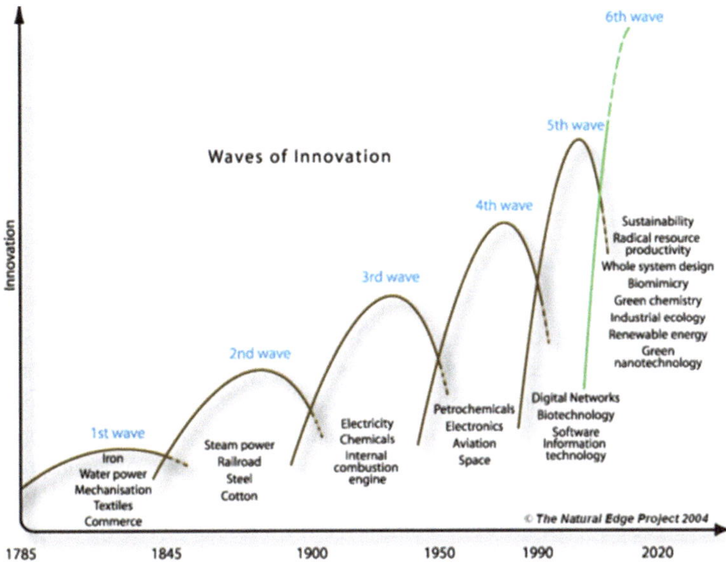

Fig. 2.2 Waves of Innovation. *Source* Hargroves (2016), with the author's permission

2.2.5 Braudel's Three Times of Human History

In his periodization of historical time Fernand Braudel was not aware of the changes in geological time, nor did he address the technical revolutions. Braudel (1969) mentions three historical times: (a) of the history of structures (*histoire de longue durée*), (b) of repetitive cycles (*histoire de conjuncture*), and (c) of events (*histoire événements*).

The first refers to "a history in which all change is slow, a history of constant repetition, ever-recurring cycles" (Braudel 1972, volume 1: 20–21). The second historical time has "slow but perceptible rhythms … One could call it *social history*, the history of groups and social groupings" with the goal of showing "how all these deep-seated forces were at work in the complex arena of warfare". The third historical time deals with "traditional history …what [was] called '*l'histoire événementielle*', that is the history of events. … a history of brief rapid, nervous fluctuations, by definition ultra-sensitive … But as such it is the most exciting of all, the richest in human interest, and also the most dangerous."

Braudel's temporal trilogy will be discussed below in a different context with structural time (*histoire de longue durée*) used to refer to the *time of changes in national and international order* due to revolutions (American, 1776; French, 1789; Soviet, 1917; Chinese, 1949) and major wars resulting in the international orders of Vienna (1815), Versailles (1919), Yalta and San Francisco (1945), and the 'new international disorder' since the end of the cold war (Brauch 2008; Holsti 1991). The time of repeated cycles (*histoire de conjuncture*) will be applied to the periods during which major elected policymakers (presidents, prime ministers, chancellors etc.) are in power, and the short time of events will focus on those (political, technical, economic) 'structure-creating' events that have fundamentally changed the existing global, international and national context or created a new one.

2.2.6 *Kondradieff's Long Cycles: Periodization of Economic History*

In his discussion of industrialization, Osterhammel (2009) noted the different interpretations between economic historians influenced by *neoclassical* or *institutional* economics and those who interpreted industrialization in Europe as an expression of a *special European way* (*europäischer Sonderweg*). He argued that most classical theories have interpreted industrialization as part of a comprehensive socio-economic transformation, e.g. Karl Marx (from 1867), Kondratieff (1925) and Schumpeter (1922/1939 [1961]), Polanyi (1944), Rostow (1960), Gerschenkron (1962), Bairoch (1963), Landes (1969) and North and Thomas (1973).

Nikolaj Kondratieff and Joseph A. Schumpeter analysed industrialization as a "cyclically structured growth process of a capitalist world economy with changing lead sectors" (Osterhammel 2009: 913–914). The first Kondratieff (1780–1830) was triggered by the steam engine and its application in the textile industry; the second Kondratieff (1830–1880) was initiated by railways and steel applied in mass transport; the third Kondratieff (1880–1930) relied on electrification and chemicals also in mass transportation; the fourth Kondratieff (1930–1970) was influenced by automobiles and petrochemicals affecting individual mobility; the fifth Kondratieff (1970–2010) dealt with information and communication technology with key applications in these two sectors and finally the sixth Kondratieff (since 2010) was inspired by environmental innovations and by nano- and biotechnology and possibly also by health care.

However, none of the theories reviewed by Osterhammel addressed the long-term environmental costs of these industrial innovations and the impacts of these human interventions in the earth system (Stern 2006, 2007) caused by the exponential increase in global fossil energy consumption since 1800 (Fig. 2.3).

2.2.7 *The Ecological Impact of the Great and Global Transformations*

Polanyi (1944) addressed the 'great transformation' of Western societies caused by industrialization and the market economy, while Buzan/Lawson (2015) discussed the *Global Transformation* during the long nineteenth century and its impact on the *Making of International Relations*, arguing that this "transformation was profound,

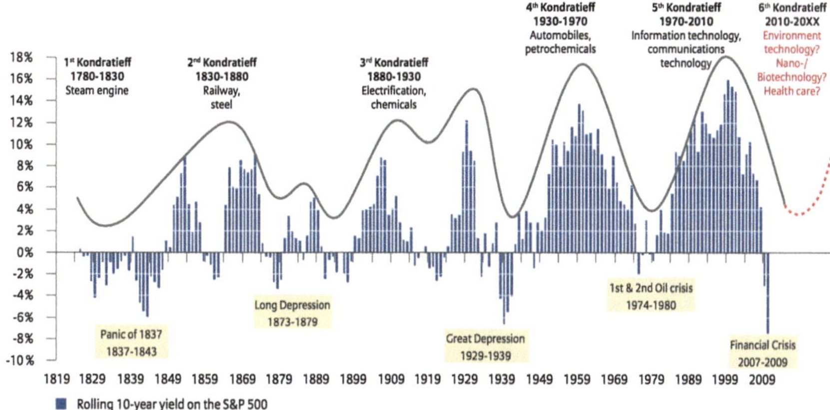

Fig. 2.3 An Interpretation of Kondratieff Waves. *Source* Naumer, Hans-Jörg; Nacken, Dennis; Scheurer, Stefan: "The sixth Kondratieff—Long waves of prosperity" (Frankfurt/M: Datastream and Allianz Global Investors Capital Market Analysis, January 2010). Permission was granted by Allianz Global Investors in August 2015

involving a complex configuration of industrialization, rational state-building and ideologies of progress" shifting from a "polycentric world with no dominant centre" to a "core-periphery order" but also "marking a shift in the distribution of power" that changed "the basic sources, or *mode of power*" (Buzan/Lawson 2015: 1).

They argue, with Osterhammel (2014: 393), that "the nineteenth century saw the birth of international relations as we know it today", since the first effect of the global transformation was "to foster the emergence of a full international system, [and] the second effect was to generate a host of new actors. Rational nation-states, transnational corporations, and standing intergovernmental and non-governmental organizations [as] leading participants in international affairs" (Buzan/Lawson 2015: 2–3). They argue that the global transformation generated four linked changes in international relations: (1) industrialization and the extension of the markets; (2) processes of rational state-formation; (3) the new ideologies, e.g. liberalism, nationalism, socialism, scientific racism; (4) this tripartite configuration fostered a core–periphery global order and destabilized relations between the great powers.

However, the change in the modern international order since the end of the Thirty Years War (1618–1648) was closely related to the emergence of the modern Westphalian state and the thinking and practice of the principles and conventions of international law (Grotius 1625). It was, however, significantly influenced by the impact of the different speeds of industrialization and the emergence of powerful war industries from the 1870s and 1890s as well as from 1940 and 1950 when the United States adopted its first global military strategy (NSC–68), which required defence expenditure to be quadrupled. Following the massive mobilization of

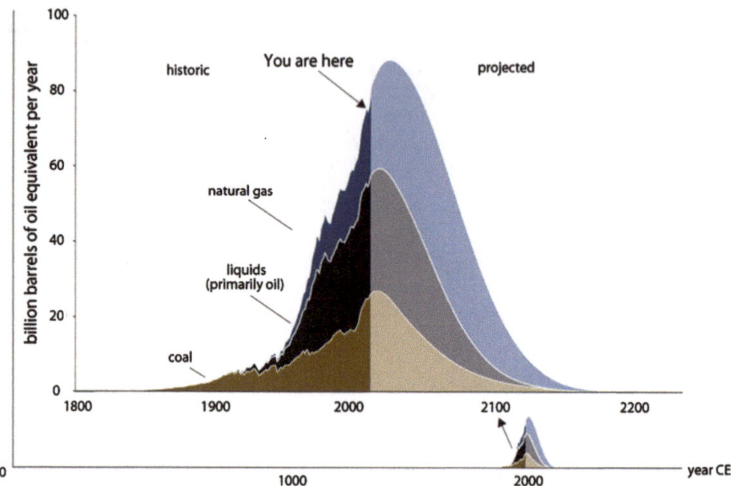

Fig. 2.4 Global production and projection of fossil fuels (1800–2200). *Source* Amory Lovins: *Reinventing Fire: Bold Business Solutions for the New Energy Era* (White River Junction, VT: Chelsea Green); at: http://www.chelseagreen.com/reinventing-fire#sthashekkRAqO0.dpuf

scientists into the war effort, the speed of innovation also partly benefitted from the spin-offs of military inventions, while in the era of *international relations* (IR) the military increasingly relies on technological break-throughs in the civil sectors. The accelerated speed of technological innovation has also resulted in an exponential increase in the consumption of fossil energy fuels (Figs. 2.1 and 2.4) and the related concentration of greenhouse gases in the atmosphere (Fig. 2.6).

2.3 International Order: Historical–Political Turning Points, Global Transformations and Transitions

From Braudel's perspective, international order is a configuration of a longer duration or an object of structural history (*histoire de longue durée*). According to Hanagan (2012), international order refers "to the structure, functioning, and nature of the international political system" but many specialists in *international relations* (IR) "disagree on how order originates and how it functions".

Since 1492, several changes in the dominant international order have occurred.

- The *Hispanic World Order*: expulsion of the Arabs and conquest of the Americas (1492–1618) by Spain and Portugal, resulting in a global order dominated by the Christian 'civilized world' that perceived the South as 'primitive barbarians';

- The *Peace of Münster and Osnabrück* (1648) after the Thirty Years War over religion (1618–1648), and the emergence of the Westphalian European order based on territorial states and an emerging international law;
- The *Utrecht Settlement* and the century of war and peace in the order of Christian princes (1715–1814).

After the independence of the United States (1776), the French Revolution (1789), the wars of liberation in Latin America (1809–1824) and the emergence of many new independent states (1817–1839), four major international orders can be distinguished. They result from

- The *Treaty of Vienna* (1815) and the European order of a balance of power based on a Concert of Europe (1815–1914) in an era of imperialism (Africa, Asia) and postcolonial liberation in Latin America.
- The *Peace of Versailles* (1919), with the collapse of the European world order, a declining imperialism and the emergence of two new power centres in the USA and the USSR with competing political, social, economic, and cultural designs; and a new global world order based on the security system of the *League of Nations* (1919–1939).

While in the first two international orders at the beginning and end of the 'global transformation' (Buzan/Lawson 2015) European powers still dominated, in the settlement of Yalta and San Francisco two new superpowers, the USA and the USSR, prevailed, while the UK and France were severely weakened:

- The *Political Settlement of Yalta* (February 1945) and the system of the United Nations discussed at the conferences in Dumbarton Oaks (1944) in Washington DC and in *Chapultepec* in Mexico City (January/February 1945), and adopted at *San Francisco* (April/June 1945) with the signing of the Charter of the United Nations.
- With the end of the cold war no new lasting international order emerged in the context of *The Political Agreement of Paris* (1990) within the framework of the *Organization for Security and Co-operation in Europe* (OSCE), which had hardly any impact on the subsequent 'new global disorder'.

Thus, in modern history four major turning points in international order were triggered by

- *the French revolution* (1789) and the Napoleonic wars that resulted in the order of Vienna (1815), where the young revolutionary republic in North America was not yet represented;
- *the Russian revolution* (1917) and *World War I* (1918) that led to the order of Versailles (1919), where although US President Woodrow Wilson had played a major role in its design, the US never joined the *League of Nations* whose structure he had helped draft;

- *World War II* (1939–1945) that resulted in the order of Yalta (1945) that ensured that the order of the Charter of the United Nations of San Francisco (1945) could never materialize; and
- *The end of the cold war* as a result of a unique peaceful transition (1989–1991) that made the deterrence system obsolete but failed to prevent 'new wars'.

The political, economic and military dominance of the US relied to a great extent on the global and regional institutions it had helped to set up at Bretton Woods (World Bank, IMF) and later the OEEC that became the OECD and NATO and the many other regional pacts (ANZUS, SEATO etc., OAS).

During the cold war (1946–1991) several phases could be identified that did not affect the institutional military structure, its strategies or the key foreign policies of both blocs:

- the *start of the cold war* from after World War II (1946/1947) until the Korean War (1950);
- the *first cold war* (1950–1963) when no substantial East–West cooperation took place;
- the *limited détente* (1963–1968) with bilateral, regional and global arms control treaties;
- the *bilateral détente* with US–Soviet arms control and confidence-building agreements that led to a multilateral détente process in Europe (CSCE, 1975) and that ended with NATO's double-track decision (INF) and the Soviet invasion of Afghanistan in December 1979;
- the *second cold war* (1980–1986); and
- the brief *second détente period* (1987–1989) that ended with the fall of the Berlin Wall and a series of peaceful changes in Eastern Europe.

During the twentieth century, three major changes in international order took place. Two (1919, 1945) were the result of global wars and the third (1989) was a peaceful change in which several factors were instrumental:

- The high military cost of the cold war, up to 30 % of GNP for the USSR, together with the lack of economic and political attractiveness of the Soviet economic and political system;
- The increasing call for disarmament in Western Europe as well as during the 1980s for human rights in Eastern Europe from dissident citizen groups who called for the implementation of the human rights guaranteed in the final Act of Helsinki of August 1975.
- The efforts of the Soviet leadership under Gorbachev to reduce the costs of their war economy and to foster more openness (*glasnost*) and domestic reforms (*perestroika*).

All these structural changes were triggered, caused or influenced by specific events that initiated a series of subsequent developments, as with the murder of the Austrian Archduke Ferdinand in June 1914 by a Bosnian Serb in Sarajevo, or the deliberate attack on Poland by Germany on 1 September 1939, or the accidental opening-up of the Berlin Wall on 9 November 1989 (Bender 1994).

However, in *geological time* in earth history, the transition from the Holocene to the Anthropocene went unnoticed until natural scientists started putting climate change on the scientific agenda from the 1970s onwards and until it became a political issue from the late 1980s onwards, and Paul J. Crutzen could claim "we are in the Anthropocene".

2.4 The Industrial Revolution: Trigger for the Silent Transition in Geological Time

Following the exponential increase in the global production and consumption of fossil fuels (Fig. 2.4) from 1800 onwards and especially since the accelerated global industrialization after World War II (Fig. 2.1), a 'silent transition' in geological time has occurred, which is why Crutzen (2002) announced that "we are now in the Anthropocene". The impact of this new social construction of reality is not yet reflected in most publications in the social sciences, political science, international relations or security, peace, environment and development studies. It is not yet well understood in the global political discourse, which fails to recognize that now "we are the threat", and that we are members of the human species that has for the first time directly interfered in the earth system.

2.4.1 Changes in CO_2 Concentration in the Atmosphere

Since the end of the glacial period some 12,000 years ago, an increase of global average temperature of about 4 °C has occurred in the Holocene. This fostered the development of advanced civilizations in the *Fertile Crescent*, in south-east and east Asia (China), in Meso-America (Mayas, Aztecs) and in South America (Incas), as a direct consequence of the Neolithic or agricultural revolution. During the Holocene, warmer periods (climate optima, e.g. during the Roman empire and the medieval period) alternated with colder periods (collapse of the Roman empire, massive movements of peoples and a thousand years later the 'Little Ice Age') (Fagan 2000, 2004; Glaser 2001, 2013; Fig. 2.5).

Figure 2.6 shows the wide fluctuations in temperature during the Holocene as well as the minor change in CO_2 concentration in the atmosphere, which oscillated between 260 *parts per million* (ppm) and 280 ppm until the start of the 'Industrial Revolution'. In the first 200 years following the start of the Industrial Revolution, CO_2 concentration in the atmosphere rose from about 279 ppm (in 1750) to 315 ppm (in 1958). However, there was a significant change from 1958 onwards;

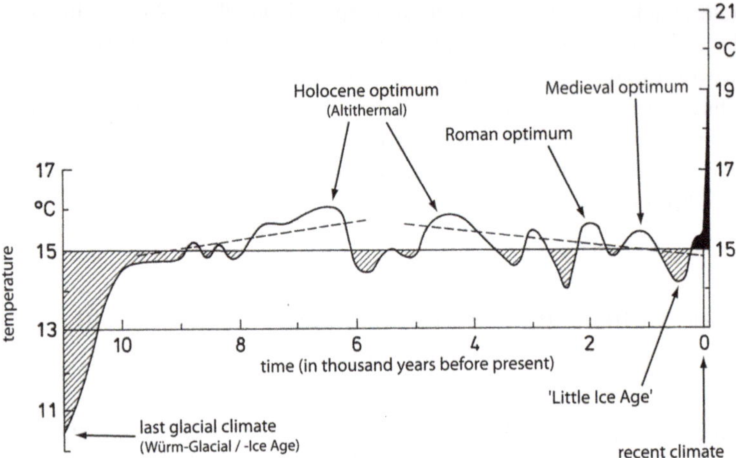

Fig. 2.5 Reconstruction of the Holocene climatic fluctuations. *Source* Blümel (2009; 104), adapted from Schönwiese (1995); reproduced with the author's permission

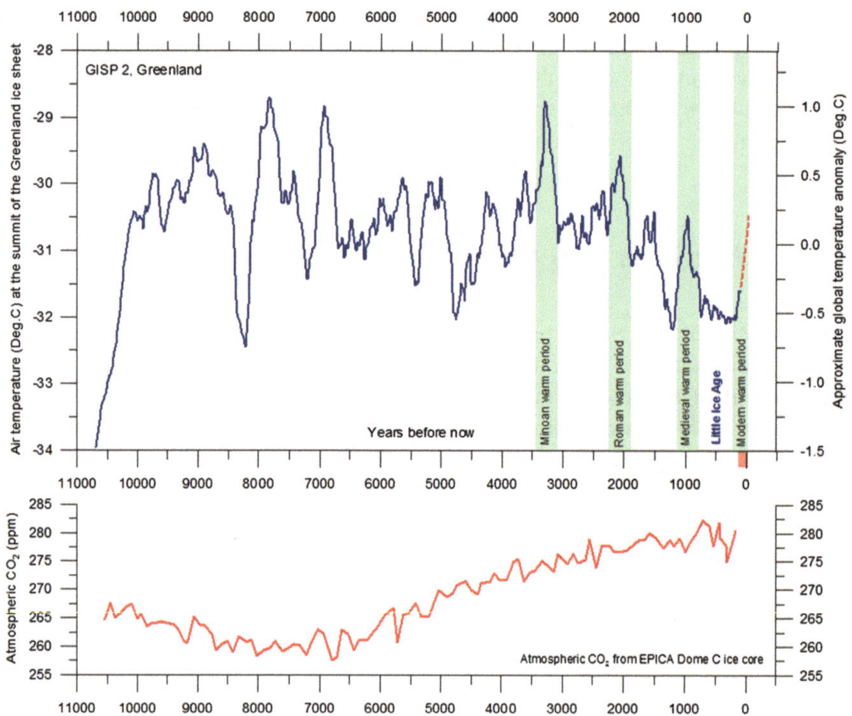

Fig. 2.6 The Holocene era of earth history. *Source* US *National Oceanic and Atmospheric Administration* (NOAA): GISP 2 and EPICA Dome C. This figure is public

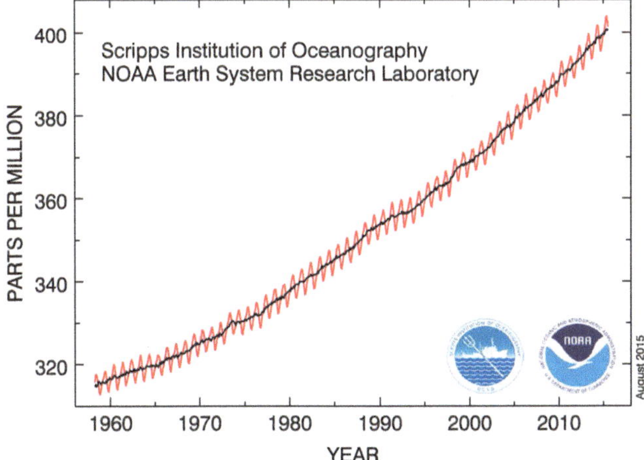

Fig. 2.7 Change in Atmospheric CO_2 at Mauna Loa Observatory. *Source* National Oceanic and Atmospheric Administration (NOAA)—Monthly Data for the Atmospheric CO_2 Since 1958 until July 2015; at: http://co2now.org/Current-CO2/CO2-Trend/ (15 August 2015). This figure is in the public domain. In June 2016 this figure rose to 406.81

by July 2015, CO_2 concentration had increased to 401.30 ppm.[13] While over 208 years (from 1750 to 1958) CO_2 concentration had risen by 36 ppm, between 1958 and July 2015 it increased by 86 ppm.

This rate of increase is more than double the increase in the 1960s (Fig. 2.7). The seasonal fluctuations from 2010 to 2014 already crossed the 400 ppm boundary.[14] The official US NOAA noted that the "atmospheric CO_2 is accelerating upward from decade to decade" and that "for the past ten years (2005–2014)", the average annual rate of increase is 2.11 parts per million (ppm). NOAA website stated that "before the Industrial Revolution in the 19th century, global average CO_2 was about 280 ppm. During the last 800,000 years, CO_2 fluctuated between about 180 ppm during ice ages and 280 ppm during interglacial warm periods. Today's rate of increase is more than 100 times faster than the increase that occurred when the last ice age ended."[15] NOAA stated that CO_2 is "increasing at an accelerating rate. … because fossil fuels are being burned at an enhanced rate, the ending of the long-term trend of increasing carbon efficiency of economies, and the ocean's diminishing absorption of CO_2" (Canadell et al. 2007).

[13]See: NOAA; at: http://www.esrl.noaa.gov/gmd/ccgg/trends/ (13 August 2015).

[14]Seasonal fluctuation in atmospheric carbon dioxide (CO_2) at the Mauna Loa Observatory. Monthly data for the Atmospheric CO_2 Since 1958. Source: National Oceanic and Atmospheric Administration (NOAA); at: http://co2now.org/Current-CO2/CO2-Trend/ (21 January 2015).

[15]See: NOAA Media Release: "Carbon Dioxide at NOAA's Mauna Loa Observatory reaches new milestone: Tops 400 ppm", 19 May 2013; at: http://www.esrl.noaa.gov/news/2013/CO2400.html.

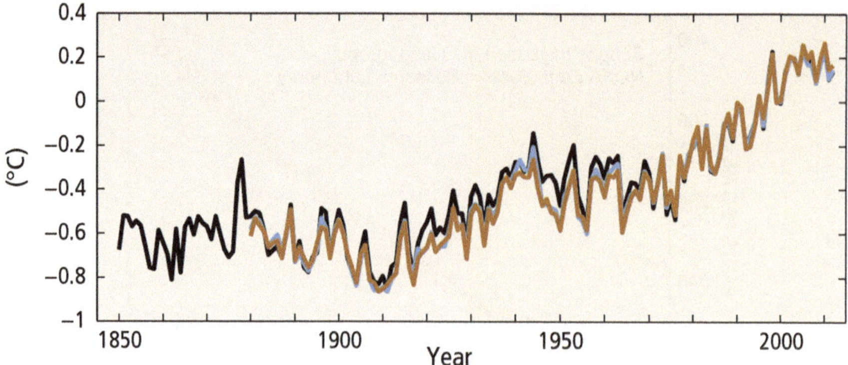

Fig. 2.8 Globally averaged combined land and ocean surface temperature anomaly (1870–2012). *Source* IPCC (2014a: 3); at: http://www.ipcc.ch/report/graphics/index.php?t=Assessment%20Reports&r=AR5%20-%20Synthesis%20Report&f=SPM. The figure is in the public domain for scientific purposes

2.4.2 *Fifth Assessment Report of the IPCC (2013/2014)*

In its Fifth Assessment Report, the *Intergovernmental Panel on Climate Change* (IPCC 2014a) summarized the observed global changes in combined land and ocean surface temperatures (1850–2012), noting that between 1900 and 2010 the global average temperature increased by about 1 °C (Fig. 2.8).

The cause has been the rapid increase since 1970, caused by combustion, in *greenhouse gases* (GHG), particularly CO_2. Despite the UNFCC and its Kyoto Protocol, this has not slowed down since 1990, although the impact of the global financial and economic crisis in 2008 had some moderating effect on GHG emissions in 2009 and 2010.[16] Depending on the chosen climate model with regard to the increase of anthropogenic CO_2 emissions, the IPCC considered an increase of more than 2 °C up to 4.5 °C above the global average temperature of 1870 as possible (Fig. 2.9).

The "Summary for Policy Makers" of the IPCC's Synthesis Report (2014a) of its Fifth Assessment stated:

- Continued emission of greenhouse gases will cause further warming and long-lasting changes in all components of the climate system, increasing the likelihood of severe, pervasive and irreversible impacts for people and ecosystems. …
- Surface temperature is projected to rise over the 21st century under all assessed emission scenarios. It is *very likely* that heat waves will occur more often and last longer, and that extreme precipitation events will become more intense and

[16]See "Total annual anthropogenic GHG emissions by gases (1970–2010)" (IPCC 2014a: 5).

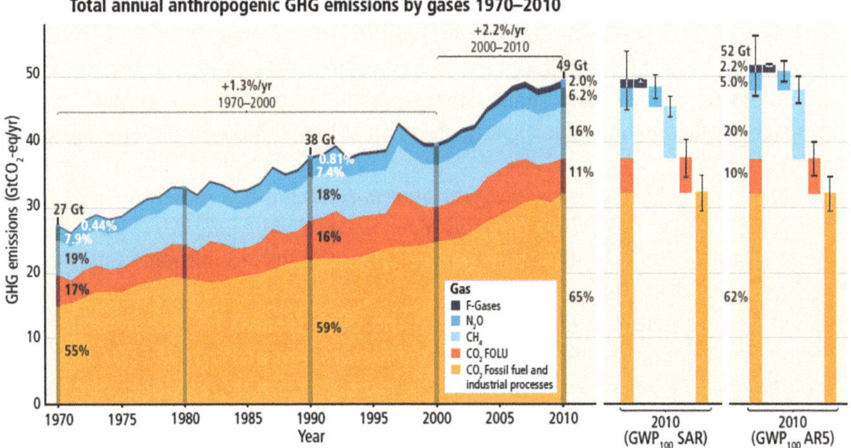

Fig. 2.9 Annual anthropogenic CO_2 emissions (1970–2010). *Source* IPCC, Synthesis Report, Summary for Policymakers (2014: 14); at: https://www.ipcc.ch/pdf/assessment-report/ar5/syr/AR5_SYR_FINAL_All_Topics.pdf. The figure is in the public domain for scientific purposes

frequent in many regions. The ocean will continue to warm and acidify, and global mean sea level to rise.

- The increase of global mean surface temperature by the end of the 21st century (2081–2100) relative to 1986–2005 is *likely* to be 0.3–1.7 °C under RCP2.6, 1.1–2.6 °C under RCP4.5, 1.4–3.1 °C under RCP6.0, and 2.6–4.8 °C under RCP8.5. The Arctic region will continue to warm more rapidly than the global mean.
- Climate change will amplify existing risks and create new risks for natural and human systems. Risks are unevenly distributed and are generally greater for disadvantaged people and communities in countries at all levels of development (IPCC 2014b: 8–14).

If the present trend of relative lack of action to reduce global GHG emissions should continue, the worst outcomes may become possible; at present, the declared goal of a 2 °C world by 2100 appears to be unlikely. This may trigger unpredictable tipping points (Lenton et al. 2008).

While so far the implications of the four key physical effects of climate change (increase in temperature, change in precipitation, rise in sea level and increase in extreme events) for security issues (migration, conflicts and wars) may have been limited (Scheffran et al. 2012; Gleditsch 2012), in its assessment of the peer-reviewed social science literature on climate change and human security the IPCC (2014b) concluded:

- *Climate change will have significant impacts on forms of migration that compromise human security (high agreement, medium evidence).* … Many vulnerable groups do not have the resources to be able to migrate to avoid the impacts of floods, storms and droughts. Models, scenarios and observations suggest that coastal inundation and loss of permafrost can lead to migration and resettlement … Migrants themselves may be vulnerable to climate change impacts in destination areas, particularly in urban centres in developing countries …

- *Mobility is a widely used strategy to maintain livelihoods in response to social and environmental changes (high agreement, medium evidence).* … Expanding opportunities for mobility can reduce vulnerability to climate change and enhance human security …
- There is insufficient evidence to judge the effectiveness of resettlement as an adaptation to climate change. *Some of the factors that increase the risk of violent conflict within states are sensitive to climate change (medium agreement, medium evidence).* The evidence on the effect of climate change and variability on violence is contested … Although there is little agreement about direct causality, low *per capita* incomes, economic contraction, and inconsistent state institutions are associated with the incidence of violence …
- *People living in places affected by violent conflict are particularly vulnerable to climate change (high agreement, medium evidence).* Evidence shows that large-scale violent conflict harms infrastructure, institutions, natural capital, social capital and livelihood opportunities. Since these assets facilitate adaptation to climate change, there are strong grounds to infer that conflict strongly influences vulnerability to climate change impacts …
- *Climate change will lead to new challenges to states and will increasingly shape both conditions of security and national security policies (medium agreement, medium evidence).* Physical aspects of climate change, such as sea level rise, extreme events and hydrologic disruptions, pose major challenges to vital transport, water, and energy infrastructure …

The data on the CO_2 concentration in the atmosphere that have been collected since 1958 at the Mauna Loa Observatory, Hawaii (Fig. 2.7) provide accepted scientific evidence that "we are in the Anthropocene!" and the fifth scientific assessment (AR5) of the IPCC confirmed that this change has been 'anthropogenic' or caused by human intervention in the earth system:

> Anthropogenic greenhouse gas emissions have increased since the pre-industrial era driven largely by economic and population growth. From 2000 to 2010 emissions were the highest in history. Historical emissions have driven atmospheric concentrations of carbon dioxide, methane and nitrous oxide to levels that are unprecedented in at least the last 800,000 years, leading to an uptake of energy by the climate system. Natural and anthropogenic substances and processes that alter the Earth's energy budget are physical drivers of climate change. Radiative forcing quantifies the perturbation of energy into the Earth system caused by these drivers. … Concentrations of carbon dioxide (CO_2), methane (CH_4) and nitrous oxide (N_2O) have all shown large increases since 1750 (40, 150 and 20 %, respectively) … CO_2 concentrations are increasing at the fastest observed decadal rate of change (2.0 ± 0.1 ppm/year) for 2002–2011 (IPCC 2014a: 44).

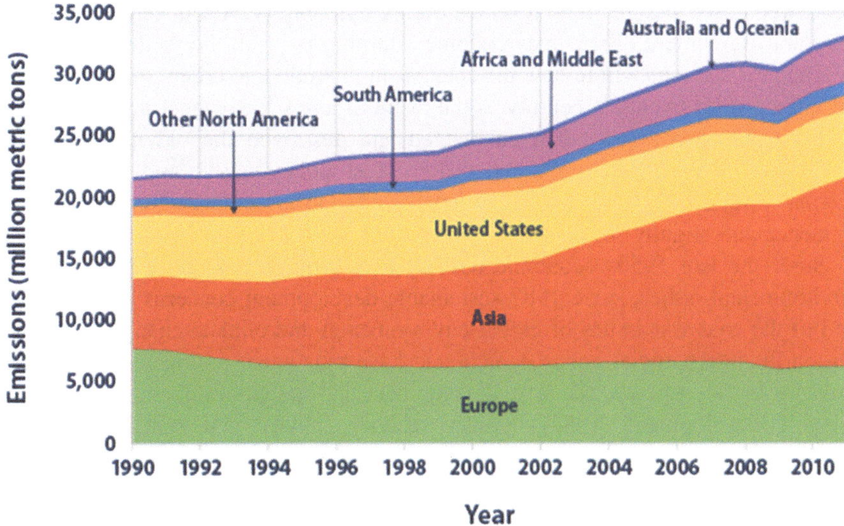

Fig. 2.10 Global carbon dioxide emissions by region (1990–2011). *Source* US EPA; at: http://www.epa.gov/climatechange/science/indicators/ghg/global-ghg-emissions.html. This figure is in the public domain

The UNFCCC (1992) counted in August 2015 195 countries and one regional organization (EU) as parties, and its *Kyoto Protocol* (KP) had 192 parties (191 countries, 1 regional organization). The US has never ratified the KP and Canada has been the only country to withdraw its ratification (in 2011). Since the reference year of 1990, global GHG and CO_2 emissions in particular have increased significantly. While most EU countries have implemented their legal commitments, many Annexe 1 countries under the UNFCCC missed their targets, among them Australia and Japan. Five OECD countries had not accepted any legally binding commitments under the KP: South Korea, Turkey, Israel, Mexico, and Chile (Fig. 2.10).

Thus, governments face a 'climate paradox' (Brauch 2012). While climate change has been accepted in numerous resolutions and declarations and by the signing of legal and political instruments, the willingness to take on obligations and to fully implement them has been mixed, and in many cases governments and parliaments have given into the massive economic pressure of the carbon lobby, including the trade unions representing workers in the coal, oil and gas industries and in the other industries that have been major emitters of GHG. In December 2015, it will become clear whether a majority of governments will be ready to accept legally binding commitments, whether their parliaments will ratify them, and whether the governments will be willing to fully implement them in close cooperation with the industries affected. The degree of implementation of a possible new climate treaty and the initiation of unilateral initiatives towards a sustainability transition will become key indicators of whether the increase in GHG emissions can be slowed down and the trend reversed during this century. It is unceretain whether the Paris Agreement (2015) may become a turning point in the Anthropocene.

The next sections will review the changes in international order (*structural time*) during the short twentieth century as the result of a few key triggering and causing events (*history of events*) that challenged and destroyed the old national order (system of rule) and international order. Most policymakers and governments (*conjunctural time*) could not fundamentally change the emerging order but only influence and slightly modify it.

Since the late 1950s economic development problems, since the early 1970s environmental issues, since 1987 sustainable development concerns and since the end of the cold war issues of global environmental and climate change have been put on the international policy agenda and have thus been 'politicized'. Gradually linkages between peace, security, development and the environment, such as sustainable development (1987), human security (1994), and sustainable peace (Brauch 2016), have been recognized and addressed by (social) scientists and policymakers and increasingly by social groups and parts of the concerned business community.

Since 2000, the challenges for peace and ecology in the Anthropocene era of earth history have gradually been recognized and addressed by environmental and peace specialists. A discussion on 'peace ecology' is just starting to emerge (Kyrou 2007; Amster 2014) and has still to be developed in the framework of the Anthropocene (Oswald Spring et al. 2014).

2.5 The Global Transformations in the Long Nineteenth Century

The *Industrial Revolution* during the long nineteenth century (technical time) brought about a *great transformation* (Polanyi 1944) of the global economic and social system and also a *global transformation* of international order (Buzan/Lawson 2015: 17–18). Buzan and Lawson focus "on how global modernity constituted a transformation" in the 'mode of power', just as the Neolithic revolution constituted a transformation based on a change in the 'mode of production'.

Both the Neolithic and the Industrial Revolutions triggered transformations: (a) a transformation in the "scale of social orders and their mode of organization" (social relations, hierarchies, social roles, collective power etc.); (b) "this configuration is produced and reproduced through inter-societal interactions" which gradually spread in space; (c) the development of new power configurations; (d) an increase in "productivity and population, plus an increase in the complexity of social orders and physical technologies, have produced a denser, more deeply connected international order" that has increased "the levels of interdependence within the international sphere" (Buzan/Lawson 2015: 19–20). These macro-transformations fostered "a new *mode of power*" that was driven by "capitalism, imperialism and ideologies of progress", but the process of spreading was uneven "as global modernity was uneven in both origins and outcomes" and has "produced larger, more complex social orders bound together in denser, more independent ways", where "industrialization produced a single world economy for the first time". While the 'Neolithic Revolution' took centuries or millennia to materialize, the 'Industrial Revolution' occurred within decades (Buzan/Lawson 2015: 21–23).

During the global transformation of the nineteenth century, "global modernity uprooted the basis on which social order rested". This transformed the local into the global and gave a few Western states an unprecedented and temporary dominance and a hegemonic position over many "aspects of international relations" (new forms of organizations, ideas etc.), with an increased economic accumulation in the countries of the industrialized centre (Bairoch 1981). In seeking to interpret why this transformation occurred, Buzan and Lawson (2015: 28–30) distinguish four accounts in the literature: (a) economic accounts from liberal or Marxist perspectives; (b) political dynamics (state formation, institutional development, role of wars and preparation for war); (c) ideational schemas (the Enlightenment, religions (Protestantism, Calvinism)); and (d) the geographical, demographic and technological advantages of the West. They conclude that "global modernity arose from a configuration of industrialization, rational state-building and ideologies of progress".

This global transformation, they argue, was a powerful "social invention" (Mann 1986: 525); the 'European miracle' was "capital-intensive, energy-intensive and land-gobbling" (Pomeranz 2000: 207) and "the uneven extension of the market through imperialism and finance capitalism generated a core-periphery order" whose fluctuations "controlled the survival chances of millions of people around the world". Industrialization was accompanied by the rational state that launched major improvements in infrastructure while ideologies (e.g. nationalism) influenced why and how wars were fought.

Buzan/Lawson (2015: 43–45) point to three impacts of the global transformation in international relations: (a) the spread of "industry, finance, railways and the telegraph, along with practices of colonialism"; (b) the development of a strong core and weak peripheral relations in trade, and (c) upheaval among the great powers (Britain, Germany, USA, Japan) who led the transformation as well as among those who tried to postpone it (China, Russia, Ottoman Empire). They concluded that "the global transformation … profoundly influenced the construction of the modern international order", setting the "material conditions under which a global international system came into being".

They analyse a classic theme of IR in a chapter on the "transformation of great powers" and "great power relations and war", and they address the implications of the global transformation due to industrialization, the rational state and ideologies with a change from 'centred' to 'decentred globalism'. But in the whole book and in their concluding chapter on "Rethinking international relations", the environmental consequences of the global transformation, including the silent transition from the Holocene to the Anthropocene, are not discussed or even noted.

2.6 The Changes of 1914 and 1919: Triggered by World War I

In 1914, the long nineteenth century ended and the short twentieth century started with World War I, when the old international order of Vienna collapsed and five years later the short-lived international order of Versailles evolved. While the order of Vienna prevented a major war in Europe that would have drawn in all the major European powers, the long nineteenth century coincided with the colonial and imperialist period, when European colonial powers (Spain, Portugal, Britain, France, Belgium, the Netherlands, Germany, Italy, Denmark) used their scientific and technological superiority to exploit the natural resources and the people of Latin America, Africa and Asia.

The United States and Japan were latecomers in extending their respective zones of economic (open-door), political and military influence. Following the Monroe Doctrine of 1823, the US intervened militarily in Latin America many times, and expelled Spain from Cuba (1898) and the Philippines (Blechman/Kaplan 1976); Japan occupied parts of China from 1931, and expanded its military influence to South East Asia during World War II.

In June 1914, the murder of Archduke Ferdinand triggered a chain of events that tumbled the Vienna order. Within a few weeks most European powers were at war, a war that resulted in total mobilization and the industrialization of and a revolution in warfare. Four years later, four empires (Russian, Austrian, German and Turkish) had collapsed and twenty million people had died.

Holsti (1991: 211) noted that "the scientific peace of 1919 created as many problems as it solved". The peace of 1815 had ignored or even suppressed the problems "of nationalism and the search for statehood", while the territorial settlement of 1919

> created economic improbabilities, security problems, restive minorities, and irredentist movements throughout Eastern Europe. Many of the armed conflicts of the 1920s and 1930s were to derive from the territorial settlements of 1919. … The German settlement greatly increased the probabilities that the peace of 1919 would be the father of later tragedy (Holsti 1991: 211–212).

The 'order of Versailles' gradually collapsed because of the actions of three revisionist powers (Japan, 1931; Italy, 1935 and Germany, 1938/1939), the self-isolation of the USA, and 'appeasement' by Britain and France. The rise of fascism in Italy and national socialism in Germany, and the systemic competition between the USA and the Soviet Union (who both remained outside the League of Nations) had prevented that the first collective security system of the League of Nations from establishing a new stable international order. With the German attack on Poland and the outbreak of World War II, the various visions of the founders of the system of Versailles became obsolete (Holsti 1991; Brauch 1996): American idealism (Woodrow Wilson), French realism (Georges Clemenceau) and British pragmatism (Lloyd George).

The political impacts of the global economic crisis of 1929 furthered the aims of the totalitarian revisionists in Japan (invasion of China) as well as the occupation of Ethiopia by the Italian Fascists, and also benefitted the Nazis in Germany, whose war economy had harshly ended the Franco-German diplomatic reconciliation of the late 1920s initiated by Gustav Stresemann and Aristide Briand; it also ended the relevance of the so-called *Kellogg-Briand Pact* (1928), which renounced war as an instrument of national policy.

Influenced by the pacifist and idealist Wilsonian tradition, peace research began to emerge in the US, in the UK and in Germany (Wright 1942), as an alternative to the geopolitical (Haushofer 1932) and geostrategic (Mahan 1890, 1897) realist tendencies in international relations and in war and peace. After World War II, there was a realist backlash in the UK and in the United States, promoted by many European immigrants (Morgenthau 1948, 1951; Herz 1959; Kissinger 1994). During the 1940s, a major transformation had taken place in the US. It had undergone a transition from isolation to active global engagement, using its economic and military power and the political will to employ these power resources in setting up the post-war world (Brauch 1977).

2.7 The Changes of 1945: Triggered by World War II

During World War II the grand coalition of the United States, Britain, the Soviet Union and later France defeated the two major revisionist powers, Germany and Japan. However, the post-war collective security system promoted by Roosevelt and Churchill during the war was paralysed by the price they had to pay Stalin at Yalta in February 1945. This was the right of the five permanent members of the Security Council of the United Nations to veto its decisions.

From 1946 the rivalry between the USA and the Soviet Union resulted in a global bipolar power structure with two military alliances (NATO vs. the Warsaw Pact). This divided not only Europe but also most regions of the world; *proxy wars* were fought in the Third World by parties who were militarily and economically supported by one or the other alliance.

While American hawks claimed that US military and economic power was instrumental in the ending of the cold war (Anderson), the collapse of the Soviet Union and the dissolution of the Warsaw Pact, others claimed that it was learning from experience, a new way of thinking by the Soviet leadership under Gorbachev (Grunberg/Risse 1992),[17] and the will of the people of Eastern Europe that brought about the first peaceful global change.

[17]Grunberg, Isabelle; Risse-Kappen, Thomas, 1992: "A Time for Reckoning? Theories of International Relations and the End of the Cold War", in: Allan, Pierre; Goldmann, Kjell (Eds.): *The End of the Cold War* (Dordrecht: Martinus Nijhoff Publishers): 104–46.

2.8 The Changes of 1989: Peaceful Change and New Wars

It was during the cold war period of 1946–1989 that global environmental issues gradually came to be seen as new political concerns, especially following the UN conference on the environment in Stockholm in 1972. Fifteen years later, the Brundtland Report called for a shift towards sustainable development and environmental security in the dominant economic path being followed. In June 1988, the Norwegian Prime Minister, Gro Harlam Brundtland, spoke of how climate change could pose a threat to security; in September 1988 the Reagan Administration put climate change on the agenda of the G7 in Toronto; in November 1988, Soviet president Gorbachev called for ecological security; and in December 1988 the UN General Assembly approved a mandate for the negotiation of a UN Treaty on Climate Change and set up the *Intergovernmental Panel on Climate Change* (IPCC).

The end of the cold war coincided with a growing awareness of new global environmental challenges that would require global cooperation. The increasing amount of new scientific knowledge on global environmental and climate change (*scientization*) was gradually perceived as a political challenge (*politicization*). Since 2004, possible links between climate change and security (*securitization*) have been addressed by the British (2004) and German governments (2007), as well as major military think tanks in the US (2007). They were placed on the agenda of the EU (2008a, b) and the UN (UNSC 2007, 2011; UNGA 2009; UNSG 2009; Brauch/Scheffran 2012; Gleditsch 2012), and research on these linkages was reviewed by the IPCC (2014b).

Three factors led to a reconceptualization of the state-centred, narrow, national concept of political and military security: the end of the cold war, the consequences of globalization, and concern over the impact of global environmental change (Brauch et al. 2008, 2009, 2011).

Unlike the situation after World War II, there was no agreed post-cold war planning for eastern and south-eastern Europe or for the rest of the world. In 1990 and 1991, the Soviet Union disintegrated, and was dissolved in late 1991, as was the Warsaw Pact. From 1991 Yugoslavia underwent a violent dissolution into five sovereign countries: Slovenia, Croatia, Bosnia-Hercegovina, Macedonia and Serbia. Kosovo (1999) left Serbia following a war, while Montenegro (2006) declared its independence following a referendum, and was recognized by Serbia.

The summit in Paris in November 1990, instead of discussing the post-cold war order of peace and security, was dominated by the US Administration's call for a war against Iraq that would end Iraq's occupation of Kuwait that had begun in summer 1990. Following an ultimatum issued to Iraq by the *United Nations Security Council* (UNSC), a war was launched in March 1991 and resulted in the defeat of Iraq.

Following the geopolitical changes in Eastern Europe in the autumn of 1989, in March 1990 Lithuania declared its independence from the Soviet Union. By the end of 1991, the Soviet Union had dissolved into its fifteen former republics. On 21

June 1991, Slovenia and Croatia declared their independence from Yugoslavia. This triggered a ten-day war in Slovenia and a bloody war between Serbia and Croatia (1991–1995), as well as in Bosnia (1992–1995) and later in Kosovo (1999). NATO, without a UN mandate, intervened twice against Serbia.

The political impacts of 1989 were mixed. The new global security environment facilitated a reunification of Germany in 1990 and the enlargement of the EU in 2004. The EU expanded to include three former Soviet republics (Estonia, Latvia and Lithuania), three former Warsaw Pact countries (Poland, Hungary and Czechoslovakia, which in 1993 had split peacefully into the Czech Republic and Slovakia), and the former Yugoslav republic of Slovenia.

A decade after the fall of the Berlin Wall, NATO gradually expanded. Poland, Hungary and the Czech Republic joined on 12 March 1999, followed on 29 March 2004 by seven additional East European countries: Bulgaria, Estonia, Latvia, Lithuania, Romania, Slovakia and Slovenia. On 1 April 2009 Albania and Croatia became NATO members and Montenegro, Bosnia-Herzegovina and Macedonia became candidates, while Kosovo, Georgia and Ukraine expressed an interest in joining NATO. Although in 2007 the parliament of Serbia expressed its military neutrality, in December 2014 the Ukrainian parliament declared its military neutrality obsolete.

South East Asia had been a battlefield for twenty-five years during the first (1950–1954) and second Indochina wars (1961–1975) that affected divided Vietnam, Laos and Cambodia. After 1990 gradual political unification has taken place within the framework of the *Association of Southeast Asian Nations* (ASEAN), and Vietnam (1995), Laos (1997), Myanmar (1997) and Cambodia (1999) became full members. Papua New Guinea has been an observer since 1976 and East Timor applied for membership in 2011.

With the attack on the World Trade Center in New York and the Pentagon in Washington on 9 September 2001, the United States was attacked on its own territory for the third time. The British had invaded from Canada in 1812, and the Japanese had attacked Pearl Harbour in 1941. This third attack was carried out by non-state, 'invisible', terrorist actors. A number of actions have contributed to a further destabilization of the Middle East: the response to 9/11 by US President George W. Bush with his 'war on terror' against al-Qaida in Afghanistan in the autumn of 2001, the attack on Iraq in 2003 that was based on falsified intelligence claims concerning Iraq's weapons of mass destruction, and the bombing of Libya by NATO air forces in March 2011 in response to a UN call for a no-fly zone, and all this has fuelled the 'hate' of radicalized Muslims around the world against the West. Israel carried out two disproportionate reprisals against Gaza in retaliation for attacks by Hamas activists in December 2008 and again in July/August 2014, killing thousands of Palestinian civilians. Since 2011, the civil war in Syria and the US withdrawal from Iraq in December 2011 have created a new military challenge to both governments and a humanitarian tragedy for religious and ethnic minorities. The challenge is posed by the jihadists of 'Islamic State' (IS), which has radicalized many young Islamists in Europe, with the resultant terrorist attacks in France in January and November 2015 and in July 2016 in Nice.

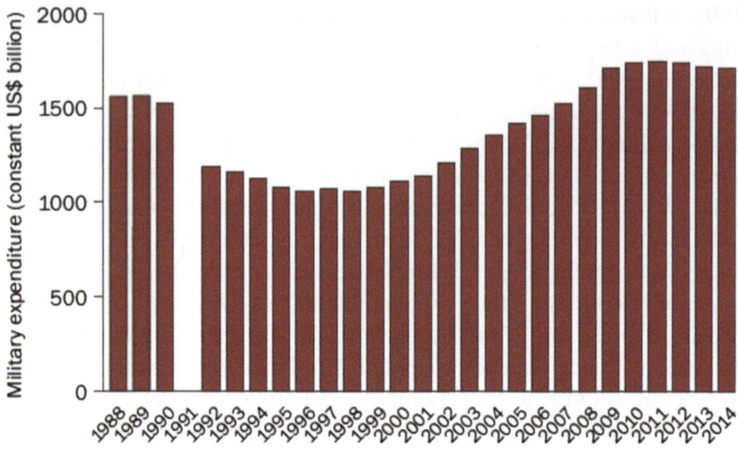

Fig. 2.11 World military expenditure (1988–2014). *Source* Stockholm International Peace Research Institute (SIPRI 2014); at: http://books.sipri.org/files/FS/SIPRIFS1504.pdf (14 August 2015). Permission to use this figure was granted by SIPRI in November 2015

Since 1990 a transformation of conflict has taken place, from the dominant bipolar rivalry, arms competition and doctrines of deterrence to 'new wars' (Holsti 1996; Kaldor/Vashee 1997; Duffield 2001), and from states to 'invisible' asymmetric non-state actors (terrorists, warlords, drug cartels, organized crime). In addition, uncontrolled greedy financial speculators have made fortunes by speculating against selected countries. This has negatively affected the livelihood and well-being of millions of poor people and fostered increasing social inequality by making the rich richer at the cost of the states and their citizens (UNDP 2014).

According to the *Stockholm International Peace Research Institute* (SIPRI 2014), the end of the cold war resulted in a temporary decline in world military expenditure between 1988 and 1996–1998, but since 1999 global military spending has risen again (Fig. 2.11), reaching a level above that at the height of the cold war. Since 2012 military expenditure in the industrialized regions has declined, while in most developing regions it has risen.[18] In 2014, nearly 80 % of global military expenditure was disbursed by fifteen states, led by the US with 37 %. The US remains the largest arms exporter. And so, twenty-five years after the end of the cold war in 1989, no 'peace dividend' has occurred globally.

The *Uppsala Conflict Data Program* (UCDP) of the Department of Peace and Conflict Research of Uppsala University has found that globally the number of wars

[18]See SIPRI (2014): "Changes of military expenditure by region (2012–2013)"; see also SIPRI Fact Sheet (April 2015): *Trends in World Military Expenditure, 2014*; at: http://books.sipri.org/files/FS/SIPRIFS1504.pdf (14 August 2015).

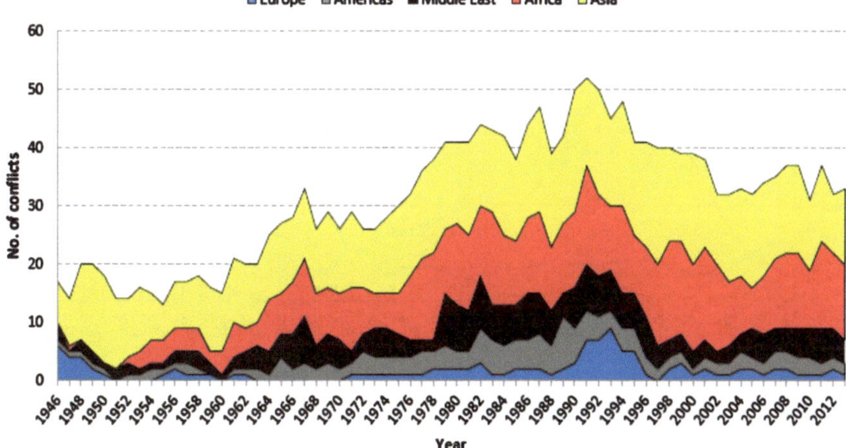

Fig. 2.12 Armed conflict by region (1946–2013). *Source* This figure was first published in an UCDP report on the UCDP website *at:* http://ucdp.uu.se/?id=1 for which the authors hold the copyright and on 26 November granted the permission to use it in this publication. This figure was later included in: Themnér et al. 2014: "*Armed Conflict, 1946–2013*", in: *Journal of Peace Research*, 51,4

and war-related deaths has declined since 1989 by intensity, by type and by region (Fig. 2.12).

While the number of major wars and intra-state conflicts fell, new asymmetric wars (Holsti 1996; Kaldor/Vashee 1998; Kaldor 1999, 2002; Duffield 2001) emerged, causing some authors to claim that "(1) the number of civil wars is increasing; (2) the intensity of battle is increasing; (3) the number of civilians displaced in civil wars is increasing; (4) the number of civilians killed in civil wars is increasing; and (5) the ratio of civilians to military personnel killed in civil wars is increasing" (Melander/Öberg/Hall, Uppsala Peace Research Papers No. 9: 3).

With regard to non-state actors, the UCDP observed that in the years between 1989 and 2013, non-state fatalities significantly increased between 1991 and 1993, and in 2013 they were higher than in 1988. The increase in the number of non-state conflicts took place primarily in Africa, where it rose from five in 1989 to thirty-five by the year 2000; in 2013 it was about twenty-eight. The number of conflicts remained below ten in Asia, but since 2007 has significantly increased in the Americas and since 2009 also in the Middle East (Fig. 2.13).

The first peaceful transition of international order in modern history has resulted neither in disarmament nor in a peace dividend. The number of inter- and intra-state wars and war fatalities has declined, but the number of non-state conflicts and non-state fatalities has remained high, especially in Africa, Asia, the Americas and the Middle East. In 2014 this was also the case in Europe as a result of the conflict between Ukraine and the Russian separatists. Thus, in the aftermath of the cold war, no lasting global or regional peace order has evolved.

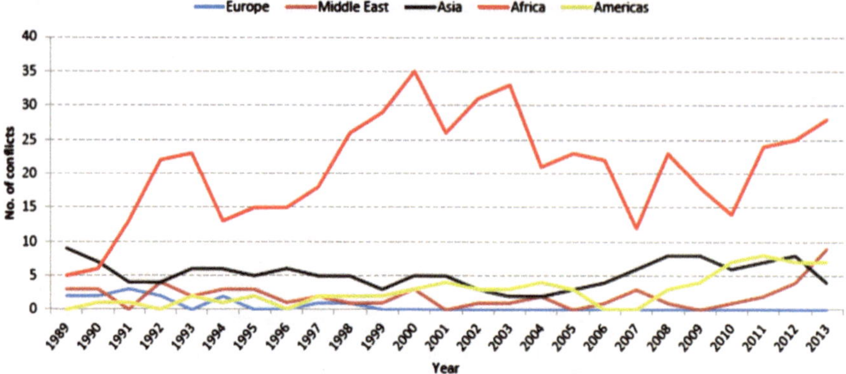

Fig. 2.13 Non-state conflicts by region (1989–2013). *Source* This figure was first published in an UCDP report ("*UCDP Non-state Conflict Dataset v. 2.5–2014 1989–2013*") on the UCDP website *at:* http://ucdp.uu.se/?id=1, for which the authors hold the copyright and on 26 November 2015 they granted the permission to use it in this publication. This figure was later included in: Sundberg et al. 2012: "Introducing the UCDP Non-*State Conflict Dataset*", in: *Journal of Peace Research*, March 2012, 49: 351–*362*

The collective security system of the UN was temporarily strengthened in 1990/1991 when the UNSC agreed to an ultimatum against Iraq. Among its regional arrangements and agencies (chapter VIII of the UN Charter) in Europe, the CSCE (which became the OSCE in 1994) extended its mandate between 1990 and 1992. But when the wars in Yugoslavia occurred in 1991, the CSCE/OSCE lacked the means to act, and in 1995 and 1999 NATO took the initiative, though without a mandate from the UNSC and in the face of opposition from Russia, and intervened militarily against Serbia. With the enlargement of NATO in Eastern Europe, NATO may have ignored political signals made by member countries towards Russia, and major governments ignored appeals made against this move by distinguished US security specialists.

Another setback occurred with the failure of international efforts to cope with the effects global environmental climate change. In the aftermath of the global financial crisis of 2008/2009, states spent billions to cope with the consequences of uncontrolled international financial markets (US stimulus plans, European efforts to counter speculation against the euro), but failed to adopt any legally-binding commitments to reduce their GHG emissions.

The international community failed to realize its policy goals under the UNFCCC (1992) and to implement its legal commitments under the Kyoto Protocol (1997). Leading democracies became climate change laggards (Fig. 2.14): Australia (+31 %), Spain (+20.2 %), Canada (+18.2 %; withdrew from the KP in 2011), Portugal (+13.1 %), Japan (+8.3 %) and the United States (+4.3 %; never ratified the KP).

Despite some changes during the first decade of the post-cold war era in the *security realm*: with a shift toward a wider and human-centred security concept, and

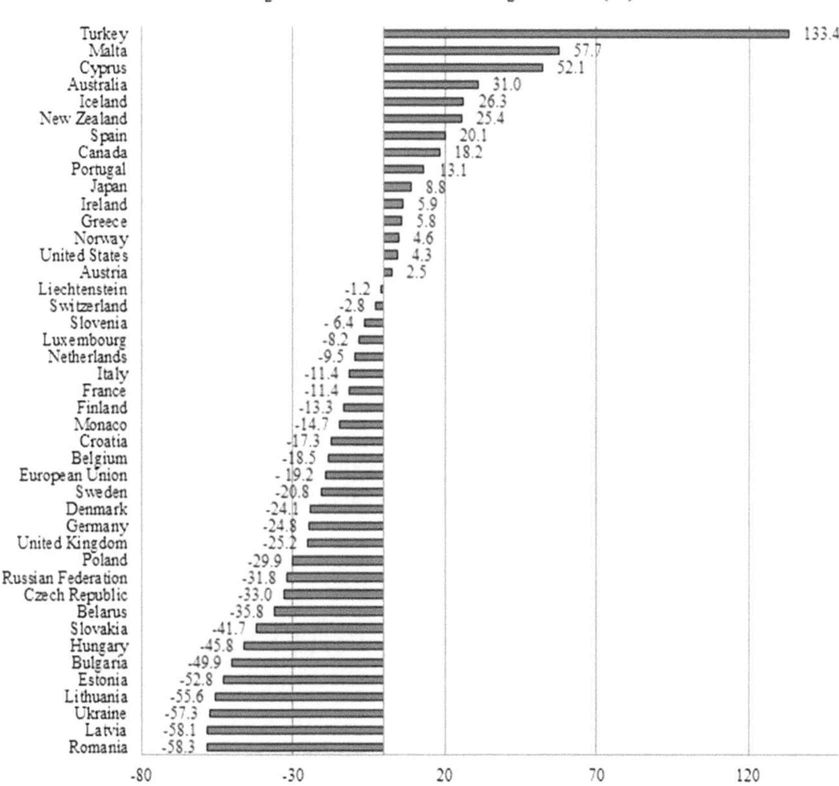

Fig. 2.14 Changes in greenhouse gas emissions excluding *Land Use, Land Use Change and Forestry* (LULUCF) from 1990 to 2012, by percentage. *Source* UNFCCC (2014), at: http://unfccc.int/files/inc/graphics/image/jpeg/total_excl_2014.jpg (24 January 2015)

in *global environment policy* with the initiative launched at the Earth Summit in Rio de Janeiro in 1992, twenty-five years after the global turn of 1989, a significant setback in both areas may be observed.

In the *security realm*, led by the US security elite and its national security apparatus, Hobbesian geostrategic thinking has returned and dominates the security policy of many countries, supported by the 'war on terror' during the Bush administration and the failure of the Obama administration to significantly revise its agenda. In some countries the military establishment has used the climate change–security nexus to legitimate its agenda, thus partially contributing to a 'militarization' of this emerging debate.

In Russia, in spring 2014, President Putin invaded, occupied and annexed the Crimea in Ukraine in clear violation of international law. Since summer 2014 in eastern Ukraine, separatist Russian forces with clandestine Russian military support have launched a new military confrontation that has challenged cooperative relations between NATO and Russia and has caused new tensions between Russia and the West, in the Baltics and in Eastern Europe.

With regard to *global environment policy*, the economically and ideologically driven policy mindset of decision-makers, and the prevailing world view in the media and in the political, economic and scientific elite that favours short-term *business as usual*, has paralysed the chance of any real changes in the Rio regimes in the fields of climate change, biodiversity loss and soil degradation and desertification, as well as the chance of any legally binding commitments at Rio+20 in June 2012. The Paris Agreement on climate change (2015) is also not legally binding.

Since the end of the cold war 'transition studies' has emerged as a discipline in political science. It focuses on the transition of state-socialist political, economic and societal systems towards Western 'neo-liberal' market economies. The emerging scientific discourse on a transition to sustainability fundamentally differs from this narrow approach; it addresses transformations in scientific, societal, economic, and political systems, as well as a radical cultural transformation that will avoid catastrophic changes in climate, soil and water as well as major losses in biodiversity.

2.9 Was 2014 a Turning Point in World History as 1914 Was?

The year 1914 was a major turning point in modern history and signalled the end of the long nineteenth century with the collapse of the international security system adopted at the Vienna Peace congress in 1814/1815. The projections of Polanyi's "Great Transformation" (1944), associated with his concept of a peaceful 'market society', had not materialized.[19] The new rationality of the 'market society' did not constrain military planners in European countries prior to 1914. Rather, World War I resulted in the industrialization of warfare, where the technological innovations were used against human beings. Prior to and during World War I two revolutions occurred, in Mexico in 1912 and in Russia in 1917. But neither these two revolutions nor the Covenant of the League of Nations could achieve the goal of a national socio-economic and political transformation and of an international security system.

The chain of events in 2014 linked to the military conflict between Ukraine and the Russian separatists in eastern Ukraine who were supported by Russia, and the role of the jihadist and terrorist group of *Islamic State* (IS) in Syria and Iraq did not trigger any longer-term structural change but strengthened mainstream thinking on national security.

---

[19]See the discussion at: https://en.wikipedia.org/wiki/The_Great_Transformation_(book).

These events may have no impact on the legally nonbinding Paris Agreement that was adopted at COP 21 of the UNFCCC in Paris in December 2015. Some longer-term commitments were offered by China and the United States in Lima (2014), and in August 2015 President Obama announced a staged transition in the energy sector from coal to renewables. It is uncertain whether such unilateral announcements and legally nonbinding obligations will be supported by the next US President and Senate after the election in November 2016 or whether the obstruction will continue that prevented the ratification of the Kyoto Protocol during the Clinton Administration (1998–2000) and that foiled all ambitious climate bills put forward by the Obama Administration (2009–2014) in the House of Representatives and in the Senate.

In 2015 policy discussions on a decoupling of economic growth from fossil energy consumption (UNEP 2011) and on a green economy (OECD)[20] are continuing and long-term policy documents issued by the European Commission (EU 2010, 2011; Happaerts 2016) are being pursued. At the G7 meeting in Elmau (Germany), the heads of the major industrialized countries stated in June 2015:

> We commit to doing our part to achieve a low-carbon global economy in the long-term including developing and deploying innovative technologies striving for a transformation of the energy sectors by 2050 … To this end we also commit to develop long term national low-carbon strategies.[21]

These stated policy goals of a decarbonization of the economy were not included in the Paris Agreement in December 2015. Whether the adopted legally nonbinding commitments will be fully implemented will be seen in the years to come. Readers will be able to judge whether COP 21 in Paris in December 2015 will initiate and reinforce a policy transition towards a decarbonization of the economy or whether short-term economic and political interests in the framework of *business as usual* will prevail. Theoretical, empirical and conceptual debates on sustainability transition (WBGU 2011) have had no influence on political agenda-setting and policy implementation.

2.10 Conclusions

This chapter has reviewed the debate in several disciplines on *cosmic*, *geological* and *technical* and by Braudel of *structural* and *conjunctural time* and of the history of events. Stimulated by Paul J. Crutzen, the key argument has been that humankind has directly interfered in the earth system since the Industrial Revolution by consuming fossil energy resources, and that this process has accelerated exponentially since 1945 with the result that "we are in the Anthropocene"; with our

[20]See publications on: "Green growth and sustainable development", at: http://www.oecd.org/greengrowth/.

[21]See G7 Leaders' Declaration, Schloss Elmau, Germany, 8 June 2015; at: https://www.whitehouse.gov/the-press-office/2015/06/08/g-7-leaders-declaration (14 August 2015).

anthropogenic interventions in the earth system we have shifted the era of geological time by moving from the Holocene to the Anthropocene era of earth and human history.

This silent transition is not yet reflected in mainstream thinking in the social sciences, international relations (Buzan/Lawson 2015), peace studies or even the early studies of peace ecology (Amster 2014). The thinking on the turning points in the short twentieth century still focuses on the events that were triggered by the two world wars (1914, 1939) and the new international orders (Versailles and Yalta). However, with the end of the cold war in 1989 and the reunification of Germany and Europe (since 1990), the causes of the 'silent transition' and the impacts of global environmental change and climate change have for the first time been put on the policy agenda.

Despite the setback in Copenhagen in December 2009 at COP 15 of the UNFCCC, and despite slow-moving and partially paralysed global climate diplomacy, the outcome of COP 21 in Paris may give global climate diplomacy a new push. While this is necessary, it will not be sufficient. What is needed is a new "scientific revolution towards sustainability" (Clark et al. 2004), and a new scientific world view, as a driver for policies whose goals are sustainable development, supported by strategies aiming at a sustainability transition (see Chap. 7 below and Brauch 2016).

References

Amster, Randall, 2014: *Peace Ecology* (Boulder, CO: Paradigm).

Bairoch, Paul, 1963: *Révolution industrielle et sous-développement* (Paris: S.E.D.E.S.).

Bairoch, Paul, 1981: "The Main Trends in National Economic Disparities Since the Industrial Revolution", in: Bairoch, Paul; Lévy-Leboyer, Maurice (Eds.): *Disparities in Economic Development since the Industrial Revolution* (London: Macmillan): 3–17.

Bauer, Franz J., 2004: *Das „lange" 19. Jahrhundert (1789–1917). Profil einer Epoche* (Stuttgart: Reclam).

Bender, Peter, 1994: "Die Öffnung der Berliner Mauer am 9. November 1989", in: Willms, Johannes (Ed.): *Der 9. November. Fünf Essays zur deutschen Geschichte* (München: C.H. Beck).

Blechman, Barry M.; Kaplan, Stephen S., 1976: *The Uses of the Armed Forces as a Political Instrument* (Washington DC: The Brookings Institution).

Blümel, Wolf Dieter, 2009: "Natural Climatic Variations in the Holocene: Past Impacts on Cultural History, Human Welfare and Crisis", in: Brauch, Hans Günter; Oswald Spring, Úrsula; Grin, John; Mesjasz, Czeslaw; Kameri-Mbote, Patricia; Behera, Navnita Chadha; Chourou, Béchir; Krummenacher, Heinz (Eds.), 2009: *Facing Global Environmental Change: Environmental, Human, Energy, Food, Health and Water Security Concepts* (Berlin–Heidelberg–New York: Springer-Verlag): 103–118.

Brauch, Hans Günter, 1977: *Struktureller Wandel und Rüstungspolitik der USA (1940–1950). Zur Weltführungsrolle und ihren innenpolitischen Bedingungen* (Ann Arbor–London: University Microfilms).

Brauch, Hans Günter, 1996: "Democracy and European Peace Order", in: *Peace Research, The Canadian Journal of Peace Studies*, 28,1 (February): 53–78.

Brauch, Hans Günter, 2002: "Climate Change, Environmental Stress and Conflict—AFES-PRESS Report for the Federal Ministry for the Environment, Nature Conservation and Nuclear Safety", in: Federal Ministry for the Environment, Nature Conservation and Nuclear Safety (Ed.): *Climate Change and Conflict. Can climate change impacts increase conflict potentials? What is the relevance of this issue for the international process on climate change?* (Berlin: Federal Ministry for the Environment, Nature Conservation and Nuclear Safety, 2002): 9–112; at: http://www.afes-press.de/pdf/Brauch_ClimateChange_BMU.pdf.

Brauch, Hans Günter, 2004: "From a Hobbesian Security to a Grotian Survival Dilemma", 40th Anniversary Conference of IPRA, Peace and Conflict in a Time of Globalisation, Sopron, Hungary, 5–9 July, at: http://www.afes-press.de/pdf/Sopron_Survival%20Dilemma.pdf.

Brauch, Hans Günter, 2008: "From a Security towards a Survival Dilemma", in: Brauch, Hans Günter; Oswald Spring, Úrsula; Mesjasz, Czeslaw; Grin, John; Dunay, Pal; Behera, Navnita Chadha; Chourou, Béchir; Kameri-Mbote, Patricia; Liotta, P. H. (Eds.): *Globalization and Environmental Challenges: Reconceptualizing Security in the 21st Century.* Hexagon Series on Human and Environmental Security and Peace, vol. 3 (Berlin–Heidelberg–New York: Springer-Verlag): 537–552.

Brauch, Hans Günter, 2009: "Securitzing Global Environmental Change", in: Brauch, Hans Günter; Oswald Spring, Úrsula; Grin, John; Mesjasz, Czeslaw; Kameri-Mbote, Patricia; Behera, Navnita Chadha; Chourou, Béchir; Krummenacher, Heinz (Eds.), 2009: *Facing Global Environmental Change: Environmental, Human, Energy, Food, Health and Water Security Concepts* (Berlin–Heidelberg–New York: Springer-Verlag): 65–102.

Brauch, Hans Günter, 2012: "Climate Paradox of the G8: legal obligations, policy declarations and implementation gap", in: *Revista Brasileira de Politica International.* Special issue edited by Eduardo Viola & Antonio Carlos Lessa on: *Global Climate Governance and Transition to a Low Carbon Economy* (Brasilia: Instituto Brasileira de Realacoes Internacionais): 30–52.

Brauch, Hans Günter, 2016: "Conceptualizing Sustainable Peace in the Anthropocene: A Challenge and Task for an Emerging Political Geoecology and Peace Ecology", in: Brauch, Hans Günter; Oswald Spring, Úrsula; Grin, John; Scheffran; Jürgen (Eds.): *Handbook on Sustainability Transition and Sustainable Peace* (Cham–Heidelberg–New York–Dordrecht–London: Springer International Publishing).

Brauch, Hans Günter; Oswald Spring, Úrsula; Mesjasz, Czeslaw; Grin, John; Dunay, Pal; Behera, Navnita Chadha; Chourou, Béchir; Kameri-Mbote, Patricia; Liotta, P. H. (Eds.), 2008: *Globalization and Environmental Challenges: Reconceptualizing Security in the 21st Century* (Berlin–Heidelberg–New York: Springer-Verlag).

Brauch, Hans Günter; Oswald Spring, Úrsula; Grin, John; Mesjasz, Czeslaw; Kameri-Mbote, Patricia; Behera, Navnita Chadha; Chourou, Béchir; Krummenacher, Heinz (Eds.), 2009: *Facing Global Environmental Change: Environmental, Human, Energy, Food, Health and Water Security Concepts* (Berlin–Heidelberg–New York: Springer-Verlag).

Brauch, Hans Günter; Oswald Spring, Úrsula; Mesjasz, Czeslaw; Grin, John; Kameri-Mbote, Patricia; Chourou, Béchir; Dunay, Pal; Birkmann, Jörn (Eds.), 2011: *Coping with Global Environmental Change, Disasters and Security—Threats, Challenges, Vulnerabilities and Risks* (Berlin–Heidelberg–New York: Springer-Verlag).

Brauch, Hans Günter; Scheffran, Jürgen, 2012: "Introduction", in: Scheffran, Jürgen; Brzoska, Michael; Brauch, Hans Günter; Link, Peter Michael; Schilling, Janpeter (Eds.): *Climate Change, Human Security and Violent Conflict: Challenges for Societal Stability* (Berlin–Heidelberg–New York: Springer-Verlag, 2012): 3–40.

Brauch, Hans Günter; Oswald Spring, Úrsula; Grin, John; Scheffran; Jürgen (Eds.): *Handbook on Sustainability Transition and Sustainable Peace* (Cham–Heidelberg–New York–Dordrecht–London: Springer International Publishing).

Braudel, Fernand 1949: *La Méditerranée et le monde méditerranéen à l'époque de Philippe II* (Paris: Armand Colin).

Braudel, Fernand, 1969: "Histoire et science sociales. La longue durée", in: *Écrits Sur l'Histoire* (Paris: Flammarion): 41–84.

Braudel, Fernand, 1972: *The Mediterranean and the Mediterranean World in the Age of Philip II*, 2 volumes (New York: Harper & Row).

Brundtland Commission (World Commission on Environment and Development), 1987: *Our Common Future. The World Commission on Environment and Development* (Oxford–New York: Oxford University Press).

Buzan, Barry; Lawson, George, 2015: *The Global Transformation: History, Modernity and the Making of International Relations* (Cambridge: Cambridge University Press).

Canadell, J. G.; Quéré, C. L.; Raupach, M. R.; Field, C. B.; Buitenhuis, E. T.; Ciais, P.; Marland, G., 2007: "Contributions to accelerating atmospheric CO_2 growth from economic activity, carbon intensity, and efficiency of natural sinks", in: *Proceedings of the National Academy of Sciences of the United States of America*, 104,47: 18866–18870.

Clapham, J. H., 1926: *An Economic History of Modern Britain: The Early Railway Age, 1820–1850* (Cambridge: Cambridge University Press).

Clapham, J. H., 1936: *The Economic Development of France and Germany 1815–1914* (Cambridge: Cambridge University Press).

Clark, William C.; Crutzen, Paul J.; Schellnhuber, Hans Joachim, 2004: "Science and Global Sustainability: Toward a New Paradigm", in: Schellnhuber, Hans Joachim; Crutzen, Paul J.; Clark, William C.; Claussen, Martin; Held, Hermann (Eds.): *Earth System Analysis for Sustainability* (Cambridge, MA; London: MIT Press): 1–28.

Crutzen, Paul J., 2002: "Geology of Mankind", in: *Nature*, 415,3 (January): 23.

Crutzen, Paul J., 2011: "The Anthropocene: A geology of mankind", in: Brauch, Hans Günter; Oswald Spring, Úrsula; Mesjasz, Czeslaw; Grin, John; Kameri-Mbote, Patricia; Chourou, Béchir; Dunay, Pal; Birkmann, Jörn (Eds.): *Coping with Global Environmental Change, Disasters and Security—Threats, Challenges, Vulnerabilities and Risks* (Berlin–Heidelberg–New York: Springer-Verlag): 3–4.

Crutzen, Paul J., Benner. Susanne; Lax, Gregor; Brauch, Hans Günter (Eds.), 2017: *Paul J. Crutzen: The Anthropocene: A New Phase of Earth History: Impacts for Science and Politics* (Cham–Heidelberg–New York–Dordrecht–London: Springer International Publishing).

Crutzen, Paul J.; Birks, John W., 1982: "The atmosphere after a nuclear war: Twilight at noon", in: *Ambio*, 11: 114–125.

Crutzen, Paul J.; Brauch, Hans Günter (Eds.), 2016: *Paul J. Crutzen: A Pioneer on Atmospheric Chemistry and Climate Change in the Anthropocene* (Cham–Heidelberg–New York–Dordrecht–London: Springer International Publishing).

Crutzen, Paul J.; Stoermer, Eugene F., 2000: "The Anthropocene", in: IGBP Newsletter, 41: 17– 18.

Dalby, Simon, 2013a: "The Geopolitics of Climate Change", in: *Political Geography*, 37: 38–47.

Dalby, Simon, 2013b: "Climate Change: New Dimensions of Environmental Security", in: *RUSI Journal*, 158,3 (June/July): 34–43.

Dalby, Simon, 2014: "Rethinking Geopolitics: Climate Security in the Anthropocene", in: *Global Policy*, 5,1: 1–9.

Dalby, Simon, 2015: "Climate Geopolitics: Securing the Global Economy", in: *International Politics*, 52,4: 426–444.

Duffield, Mark, 2001. *Global Governance and the New Wars: The Merging of Development and Security* (London: Zed).

EU (European Commission; Council), 2008a: *Climate Change and International Security*. Doc 7249/08 (Brussels: European Commission, 14 March).

EU (European Council), 2008b: *Report on the Implementation of the European Security Strategy —Providing Security in a Changing World*, S407/08 (Brussels: European Council, 11 December).

EU (European Commission), 2010: *Roadmap to a Resource Efficient Europe (Impact Assessment roadmap)* (Brussels: European Commission).

EU (European Commission), 2011: *Roadmap to a Resource Efficient Europe* (Brussels: European Commission).

Fagan, Brian, 2000, 2002: *The Little Ice Age. How Climate Made History 1300–1850* (New York: Basic Books).

Fagan, Brian, 2004: *The Long Summer. How Climate Changed Civilization* (New York: Basic Books—London: W. Clowes Ltd.).

Gerschenkron, Alexander, 1962: *Economic Backwardness in Historical Perspective* (Cambridge, MA: Harvard University Press).

Glaser, Rüdiger, 2001: *Klimageschichte Mitteleuropas—1000 Jahre Wetter, Klima, Katastrophen* (Darmstadt: Wissenschaftliche Buchgesellschaft).

Glaser, Rüdiger, 2013: *Klimageschichte Mitteleuropas—1200 Jahre Wetter, Klima, Katastrophen* (Darmstadt: Wissenschaftliche Buchgesellschaft).

Gleditsch, Nils-Petter, 2012: "Whither the weather? Climate change and conflict", in: *Journal of Peace Research, special issue: Climate change and conflict,* 49,1 (January–February): 9–18.

Grin, John; Rotmans, Jan; Schot, Johan, 2010: *Transitions to Sustainable Development. New Directions in the Study of Long Term Transformative Change* (New York, NY–London: Routledge).

Gronenborn, Detlef; Terberger, Thomas, 2014: "Die ersten Bauern in Mitteleuropa—eine interdisziplinäre Herausforderung", in: *Archäologie in Deutschland,* special issue 05/2014: *Vom Jäger und Sammler zum Bauern. Die Neolithische Revolution* (Darmstadt: Wissenschaftliche Buchgesellschaft): 7–14.

Grotius, Hugo, 1625, 1646: *De Jure Belli ac Pacis* (Amsterdam: Iohanem Blaeu).

Grotius, Hugo, ^{2}1990 [1625]: "Prolegomena to the Law of War and Peace" in: Vasquez, J.A. (Ed.): *Classics of International Relations* (Englewood Cliffs, NJ: Prentice Hall).

Gruenewald, David A, 2003: "The best of both worlds: a critical pedagogy of place," in: *Educational Researcher*, 32, 4: 3–12.

Grunberg, Isabelle; Risse-Kappen, Thomas, 1992: "A Time for Reckoning? Theories of International Relations and the End of the Cold War", in: Allan, Pierre; Goldmann, Kjell (Eds.): *The End of the Cold War* (Dordrecht: Martinus Nijhoff Publishers): 104–146.

Hanagan, Deborah L., "International order", in: US Army War College Guide to National Security Issues (Ed.): *US Army War College Guide to National Security Issues*, Volume 2: *National Security Policy and Strategy* (Carlisle Barracks, PA: Strategic Studies Institute of the US Army War College, June 2012).

Happaerts, Sander, 2016: "Discourse and Practice of Transitions in International Policy-making on Resource Efficiency in the EU", in: Brauch, Hans Günter; Oswald Spring, Úrsula; Grin, John; Scheffran; Jürgen (Eds.): *Handbook on Sustainability Transition and Sustainable Peace* (Cham–Heidelberg–New York–Dordrecht–London: Springer, *forthcoming*).

Hargroves, Karlson 'Charlie', 2016: "Considering a Structural Adjustment Approach to the Low Carbon Transition", in: Brauch, Hans Günter; Oswald Spring, Úrsula; Grin, John; Scheffran; Jürgen (Eds.): *Handbook on Sustainability Transition and Sustainable Peace* (Cham–Heidelberg–New York–Dordrecht–London: Springer, *forthcoming*).

Harlan, Jack R., 1992: *Crops & Man: Views on Agricultural Origins* (Madison, WI: American Society of Agronomy and Crop Science Society of America).

Haushofer, Karl, 1932: *Jenseits der Großmächte* (Leipzig-Berlin: B. G. Teubner).

Heidegger, Martin, 1927, ^{19}2006: *Sein und Zeit* (Tübingen: Niemeyer).

Herz, John H., 1959, 1962, 1966: *International Politics in the Atomic Age* (New York: Columbia University Press).

Hobsbawm, Eric, 1962: *The Age of Revolution: Europe 1789–1848* (London: Abacus).

Hobsbawm, Eric, 1975: *The Age of Capital: 1848–1875* (London: Weidenfeld & Nicolson).

Hobsbawm, Eric, 1987: *The Age of Empire: 1875–1914* (London: Weidenfeld & Nicolson).

Hobsbawm, Eric, 1994: *The Age of Extremes: The Short Twentieth Century, 1914–1991* (London: Michael Joseph).

Holsti, Kalevi J., 1991: *Peace and War: Armed Conflicts and International Order 1648–1989* (Cambridge: Cambridge University Press).

Holsti, Kalevi, 1996: *The State War and the State of War* (Cambridge: Cambridge University Press).

IPCC, 1990: *Climate Change. The IPCC Impacts Assessment* (Geneva: WMO; UNEP; IPCC).

IPCC, 1995: *IPCC Second Assessment Climate Change 1995. A Report of the Intergovernmental Panel on Climate Change* (Geneva: WMO, UNEP).

IPCC, 2001: *Climate Change 2001: Synthesis Report. A Contribution of Working Groups I, II, and III to the Third Assessment Report of the Intergovernmental Panel on Climate Change* (Cambridge: Cambridge University Press).

IPCC, 2007: *Climate Change 2007. Synthesis Report* (Geneva: IPCC); at: http://www.ipcc.ch/pdf/assessment-report/ar4/syr/ar4_syr.pdf.

IPCC, 2014a: *Climate Change 2014—Synthesis Report, Summary for Policymakers* (Geneva: IPCC); at: https://www.ipcc.ch/report/ar5/syr/.

IPCC, 2014b: "Human Security", in: *Climate Change 2014—Impacts, Adaptation, and Vulnerability—Part A: Global and Sectoral Aspects. Working Group II Contribution to the Fifth Assessment Report of the Intergovernmental Panel on Climate Change* (Cambridge–New York: Cambridge University Press): 755–801; at: http://www.ipcc.ch/pdf/assessment-report/ar5/wg2/WGIIAR5-Chap12_FINAL.pdf.

Jochum, Uwe, 2010: "Wissensrevolution", in: *WBG Weltgeschichte*, vol. V: *Entstehung der Moderne 1700 bis 1914* (Darmstadt: Wissenschaftliche Buchgesellschaft): 150–194.

Kaldor, Mary, 1999: *New and Old Wars: Organized Violence in a Global Era* (Cambridge: Polity–Stanford: Stanford University Press).

Kaldor, Mary, 2002: *New and Old Wars: Organized Violence in a Global Era.* (Cambridge: Polity).

Kaldor, Mary; Vashee, Basker (Eds.), 1997: *New Wars* (London: Pinter).

Kissinger, Henry A., 1994: *Diplomacy* (New York: Simon & Schuster).

Kocka, Jürgen, [10]2002: *Das lange 19. Jahrhundert. Arbeit, Nation und bürgerliche Gesellschaft* (Stuttgart: Klett-Cotta).

Kondratieff, Nikolaj, 1925, 1984: *Long Wave Cycle* (New York, NY: E. P. Dutton).

Koselleck, Reinhart, 2000: *Zeitschichten—Studien zur Historik* (Frankfurt am Main: Suhrkamp).

Kuhn, Thomas, 1962: *The Structure of Scientific Revolutions* (Chicago: University of Chicago Press).

Kyrou, Christos N., 2007: "Peace Ecology: An Emerging Paradigm in Peace Studies", in: *The International Journal of Peace Studies,* 12,2 (Spring/Summer): 73–92.

Landes, David S., 1969: *The Unbound Prometheus: Technological Change and Industrial Development in Western Europe from 1750 to the Present* (Cambridge: Cambridge University Press).

Lenton, Timothy; Held, Hermann; Kriegler, Elmar; Hall, Jim W.; Lucht, Wolfgang; Ramstorf, Stefan; Schellnhuber, Hans Joachim, 2008: "Tipping elements in the Earth's climate system", in: *Proceedings of the National Academy of Science* (PNAS), 105,6 (12 February): 1786–1793.

Mackay, A. W.; Battarbee, R. W.; Birks, H. J. B. et al. (Eds.), 2003: *Global change in the Holocene* (London: Arnold).

Mahan, Alfred, 1890: *The Influence of Sea Power Upon History, 1660–1783* (Boston: Little Brown).

Mahan, Alfred, 1897: *The Interest of America in Sea Power, Present and Future* (London: Sampson Law).

Mann, Michael, 1986: *The Sources of Social Power,* vol. 1: *A History of Power from the Beginning to AD 1760* (Cambridge: Cambridge University Press).

Melander, Erik; Öberg, Magnus; Hall, Jonathan, n.d: *The 'New Wars' Debate Revisited: An Empirical Evaluation of the Atrociousness of 'New Wars'.* Uppsala Peace Research Papers No. 9: 3 (Uppsala: Uppsala University, Department of Peace and Conflict Research).

Morgenthau, Hans J., [1]1948, [3]1960, 1961, 1969, [5]1973: *Politics Among Nations. The Struggle for Power and Peace* (New York: Alfred A. Knopf).

Morgenthau, Hans J., 1951: *In Defense of the National Interest: A Critical Examination of American Foreign Policy* (New York: Knopf).

Münkler, Herfried, 2004: *Die neuen Kriege* (Reinbek: Rowohlt).

North, Douglas C.; Thomas, Robert Paul, 1973: *The Rise of the Western World: A New Economic History* (Cambridge: Cambridge University Press).

Osterhammel, Jürgen, 2009: *Die Verwandlung der Welt. Eine Geschichte des 19. Jahrhunderts* (München: C.H. Beck).

Osterhammel, Jürgen, 2014: *The Transformation of the World: A Global History of the Nineteenth Century* (Princeton: Princeton University Press).

Oswald Spring, Úrsula, 2016: "Development with Sustainable-Engendered Peace: A Challenge during the Anthropocene", in: Brauch, Hans Günter; Oswald Spring, Úrsula; Grin, John; Scheffran; Jürgen (Eds.): *Handbook on Sustainability Transition and Sustainable Peace* (Cham–Heidelberg–New York-Dordrecht–London: Springer, *forthcoming*).

Oswald Spring, Úrsula; Brauch, Hans Günter; Tidball, Keith G., 2014: "Expanding Peace Ecology: Peace, Security, Sustainability, Equity and Gender", in: Oswald Spring, Úrsula; Brauch, Hans Günter; Tidball, Keith G. (Eds.): *Expanding Peace Ecology: Peace, Security, Sustainability, Equity and Gender—Perspectives of IPRA's Ecology and Peace Commission* (Cham–Heidelberg–New York-Dordrecht–London: Springer): 1–32.

Polanyi, Karl, 1944: *Great Transformation: The Political and Economic Origins of our Time* (Boston, MA: Beacon Press).

Pomeranz, Kenneth, 2000: *The Great Divergence* (Princeton: Princeton University Press).

Reychler, Luc, 2015a: *Time for peace: The essential role of time in conflict and peace processes* (St Lucia, Queensland, Australia: University of Queensland Press).

Roberts, Neil, 1998: *The Holocene: an environmental history* (2nd ed.) (Malden, MA: Blackwell).

Rostow, Walt W., 1960: *The Stages of Economic Growth—A Non-Communist Manifesto* (Cambridge: Cambridge University Press).

Scheffran, Jürgen; Brzoska, Michael; Brauch, Hans Günter; Link, Peter Michael; Schilling, Janpeter (Eds.), 2012: *Climate Change, Human Security and Violent Conflict: Challenges for Societal Stability* (Berlin–Heidelberg–New York: Springer-Verlag).

Schellnhuber, Hans Joachim; Cramer, Wolfgang; Nakicenovic, Nebojsa; Wigley, Tom; Yohe, Gary (Eds.), 2006: *Avoiding Dangerous Climate Change* (Cambridge: Cambridge University Press).

Schellnhuber, Hans Joachim; Hare, William L.; Serdeczny, Olivia; Adams, Sophie; Coumou, Dim; Frieler, Katja; Marin, Maria; Otto, Ilona M.; Perrette, Mahé; Robinson, Alexander; Rocha, Marcia; Schaeffer, Michiel; Schewe, Jacob; Wang, Xiaoxi; Warszawski, Lila, 2012: *Turn Down the Heat: Why a 4 °C Warmer World Must Be Avoided* (Washington DC: The World Bank).

Schellnhuber, Hans Joachim; Serdeczny, Olivia Maria; Adams, Sophie; Köhler, Claudia; Otto, Ilona Magdalena; Schleussner, Carl-Friedrich, 2016: "The Challenge of a 4 °C World by 2100", in: Brauch, Hans Günter; Oswald Spring, Úrsula; Grin, John; Scheffran; Jürgen (Eds.): *Handbook on Sustainability Transition and Sustainable Peace* (Cham–Heidelberg–New York-Dordrecht–London: Springer, *forthcoming*).

Schmidt, Klaus, 2009: "Von den ersten Dörfern zu frühurbanen Strukturen", in: Jockenhövel, Albrecht (Ed.): *WBG Weltgeschichte.* Vol. 1: *Grundlagen der globalen Welt—Vom Beginn bis 1200 v. Chr.* (Darmstadt: Wissenschaftliche Buchgesellschaft): 128–144.

Schönwiese, Christian, 1995: *Klimaänderungen—Daten, Analysen, Prognosen* (Berlin–Heidelberg: Springer).

Schumpeter, Joseph A., 1961: *Konjunkturzyklen. Eine theoretische, historische und statistische Analyse des kapitalistischen Prozesses* (Göttingen: Vandenhoeck & Ruprecht).

SIPRI, 2014: *SIPRI Yearbook 2014. Armaments, Disarmament and International Security* (Oxford: Oxford University Press).

Sirocko, Frank, [3]2012: *Wetter, Klima, Menschheitsentwicklung. Von der Eiszeit bis ins 21. Jahrhundert* (Darmstadt: Wissenschaftliche Buchgesellschaft).

Steffen, Will; McNeill, John; Crutzen, Paul J., 2007: "The Anthropocene: Are Humans Now Overwhelming the Great Forces of Nature?". in: *Ambio*, 36: 614–621.

Steffen, Will; Grinevald, Jacques; Crutzen, Paul J.; McNeill, John, 2011: "The Anthropocene: conceptual and historical perspectives", in: *Phil. Trans. R. Soc. A* 369: 843.

Stern, Nicholas, 2006, 2007, [4]2008: *The Economics of Climate Change—The Stern Review* (Cambridge–New York: Cambridge University Press).

Straßheim, Holger; Ulbricht, Tom (Eds.) 2015: *Zeit der Politik—Demokratisches Regieren in einer beschleunigten Welt*—Leviathan Sonderband 30/2015 (Baden-Baden: Nomos).

Sundberg, Ralph; Eck, Kristine; Kreutz, Joakim, 2012: "Introducing the UCDP Non-State Conflict Dataset", in: *Journal of Peace Research*, 49 (March): 351–362.

Themnér, Lotta; Wallensteen, Peter, 2014: "Armed Conflict, 1946–2013", in: *Journal of Peace Research*, 51,4: *697–710.*

UN, 2009: *Climate change and its possible security implications. Report of the Secretary-General.* A/64/350 of 11 September 2009 (New York: United Nations).

UNDP, 2014: "Inequality-adjusted Human Development Index (IHDI)", at: http://hdr.undp.org/en/content/inequality-adjusted-human-development-index-ihdi.

UNEP, 2011: *Green Economy Report: Towards a Green Economy: Pathways to Sustainable Development and Poverty Eradication* (Nairobi: UNEP); at: http://www.ipu.org/splz-e/rio+20/rpt-unep.pdf.

UNEP, 2014: *Decoupling 2: Technologies, Opportunities and Policy Options- Report to UNEP's International Resource Panel* (Nairobi: UNEP).

UNGA, 2009: "Climate change and its possible security implications". Resolution adopted by the General Assembly, A/RES/63/281 (New York: United Nations General Assembly, 11 June).

UNSC, 2007: "Security Council Holds First-Ever Debate on Impact of Climate Change on Peace, Security, Hearing over 50 Speakers, UN Security Council, 5663rd Meeting, 17 April 2007"; at: http://www.un.org/News/Press/docs/2007/sc9000.doc.htm.

UNSC, 2011: "Statement by the President of the Security Council on "Maintenance of Peace and Security: Impact of Climate Change", S/PRST/1011/15, 20 July 2011.

UNSG, 2009: "Climate Change and its Possible Security Implications" of 11 September 2009 (A/64/350).

WBGU, 2011: *World in Transition—A Social Contract for Sustainability* (Berlin: German Advisory Council on Global Change, July 2011).

WCED (World Commission on Environment and Development), 1987: *Our Common Future. The World Commission on Environment and Development* (Oxford–New York: Oxford University Press).

Wright, Quincy, 1942, 1965: *A Study of War* (Chicago–London: University of Chicago Press).

Zalasiewicz, Jan; Williams, Mark; Smith, Alan et al., 2008: "Are we now living in the Anthropocene?", in: *GSA Today*, 18,2: 4–8.

Zalasiewicz, Jan; Williams, Mark; Steffen, Will; Crutzen, Paul, 2010: "The New World of the Anthropocene", in: *Environment Science & Technology*, 44,7: 2228–2231.

Ziegler, Dieter, 2009: "Die Industrialisierung", in: *WBG Weltgeschichte*, vol. V: *Entstehung der Moderne 1700 bis 1914* (Darmstadt: Wissenschaftliche Buchgesellschaft): 41–91.

Zimmermann, Andreas, 2009: "Neolithisierung und frühe soziale Gebilde", in: Jockenhövel, Albrecht (Ed.): *WBG Weltgeschichte.* Vol. 1: *Grundlagen der globalen Welt—Vom Beginn bis 1200 v. Chr.* (Darmstadt: Wissenschaftliche Buchgesellschaft): 95–127.

Chapter 3
Global Ecological Crisis: Structural Violence and the Tyranny of Small Decisions

Juliet Bennett

Abstract What is causing the global ecological crisis? Who has the power to solve it? This chapter explores the global ecological crisis as a form of structural violence. Galtung's "Structural Theory of Imperialism" (1971) is integrated with Kahn's "Tyranny of Small Decisions" (1966). The synthesis of theories sheds light on the multi-levelled and multi-directional influence of individuals, nations, institutions and culture. Countless "small decisions", that appear separate and distant from their collective long-term global consequences, are posited to be a root cause of the crisis. Solving the crisis calls for a holistic re-orienting of decision-making by people across many sectors of society aimed at long-term global interests rather than short-term personal interests. Examples of these decisions are considered. The chapter closes by imagining what a just and sustainable world system operating within planetary boundaries might look like, and consider examples of the type of decision-making it might involve.

Keywords Global ecological crisis · Structural violence · Tyranny of small decisions · New story · Holistic worldview · Process philosophy

3.1 Introduction

What is causing the global ecological crisis? Who has the power to solve it? What can motivate them to do so? These questions are of utmost importance for shaping a peaceful or violent future for humanity and other species. They are also, of course, too big and complex for one person or one chapter to answer. This chapter offers an introduction to the complex relationships between politics, economics, culture and

Juliet Bennett is a Ph.D. Candidate at the Centre for Peace and Conflict Studies, The University of Sydney, Australia; Email: juliet.bennett@sydney.edu.au. The author is grateful for the comments and suggestions of the anonymous reviewers on a prior draft of this chapter, as well for the feedback and support of Stuart Rees, Garry Trompf and Lynda Ann Blanchard on her M.Phil thesis.

H.G. Brauch et al. (eds.), *Addressing Global Environmental Challenges from a Peace Ecology Perspective*, The Anthropocene: Politik—Economics—Society—Science 4, DOI 10.1007/978-3-319-30990-3_3

ecology as viewed from an interdisciplinary peace and conflict studies perspective, in hope of shedding light on these two questions.

The central argument is that the global ecological crisis is a form of structural violence, an indirect form of violence for which no one is directly responsible (Galtung 1969). The chapter posits that the crisis has resulted as an unintended consequence of countless everyday decisions by individuals in their roles within institutions and nations. These decisions are influenced and limited by historically embedded macro-structures (such as policies, laws and cultural norms). Arguably, however, the collective decision making of individuals has the power to evolve those structures to be more just and sustainable. In other words, it is neither solely the structures nor solely the actors who are responsible for the global ecological crisis, but it is the *interaction* between them.

The argument will unfold in three stages. First, Sect. 3.2 clarifies what the author is referring to by 'global ecological crisis' and introduces some of the complexities around its causes. This stage surveys some of the well-known dimensions of the global ecological crisis such as climate change and loss of biodiversity, in a broader context particularly focused on a paradox between population stabilization, entrenched poverty, a rampant profit motive and a planetary ecosystem with limits. This feeds into the next stage, which seeks to answer the second research question: Who has the power to address the crisis?

Section 3.3 brings together a number of theories and examples that help to explain the global ecological crisis as a form of structural violence, and to point to varying power of people and institutions to mitigate it. Galtung's widely cited "A Structural Theory of Imperialism" (1971) is selected as an example of dependency and world systems theories, providing a critical perspective of the global distribution of political and economic power. This model is expanded with reference to Jorgenson (2006), to propose that this imperialist structure continues to influence unequal ecological exchanges between higher and lower income countries,[1] and is an obstacle to successful international climate change negotiations.

The power of individual actors within this model is then considered with reference to Kahn's economic theory "Tyranny of Small Decisions" (1966), observing the "small" nature of decisions, short-term and locally focused, that are inadvertently causing the global crisis. The model is further expanded with reference to Sklair (2002) to propose that power is particularly concentrated in a Transnational Capitalist Class, a power network of corporate, government, professional, media and consumer elites. As a whole the theories and their synthesis offers an introduction to some of the global political, economic and cultural factors, and groups of actors, which have contributed to the global ecological crisis and have some power to mitigate it.

[1]The language of high-income countries and low-income countries is preferred to the corresponding first world and third world, Global North and Global South, or developed and developing worlds, however these terms will be used interchangeably due to differing terms used in the literature reviewed.

The final stage of the argument explores the ways in which a holistic re-orienting of decisions by groups of actors might work to bring about a more just and ecologically sustainable world system. Inspired by the workshops of Boulding (1988), the author indulges an imaginary leap into what such a system might look like. This is supported by intersecting discourse in process philosophy, deep ecology and macro history aimed at moving towards an ecological civilization. Examples are provided to consider to the types of decision-making that such a shift might involve.

Before exploring the dynamics of the structural violence, it will be valuable to clarify exactly what is being referred to here as a 'global ecological crisis.'

3.2 Global Ecological Crisis

Our foul air, polluted waters and oceans, shrinking croplands, creeping deserts and extinguishing species tell the true story (McDonagh 1986: 45).

Seminal works such as Leopold's philosophy of a "Land Ethic" in *A Sand County Almanac* (1949), Carson's *Silent Spring* (1965), *Limits to Growth* (Meadows et al. 1972), and *The Economist's Blueprint for Survival* (1972) by Goldsmith and Allen, have gradually increased awareness and concern about the effect that humans are having on their environment. Thanks to countless books and documentaries such as former Vice-President Gore's *An Inconvenient Truth* (2006), awareness and concern for humanity's impact on the environment now has widespread public awareness. As Sean McDonagh points out in the quote above, these observations tell the true story. This section reviews some of the key issues in order to clarify what is the 'global ecological crisis' and how it has arisen.

Signs of a global ecological crisis include air pollution, climate change, the vast loss of millions of species, loss of biodiversity and topsoil, overgrazing and disruptive agricultural practices, related issues of desertification and deforestation, and disrupted water systems (Rajagopalan 2011). Humanity is witnessing a "systemic destruction" of nature, species, societies and cultures, which poses a potential threat to the very "survival of biological life" (Escobar 1997). Some even call it ecocide and suggest that it should be internationally recognized as "the 5th Crime against Peace" (Higgins et al. 2013). Rockström and his colleagues (2009) have identified "planetary boundaries" as a framework for humanity to limit the impact of their activities on the planet.

If humanity is going to avoid disasterous and violent consequences, they must stay within nine planetary limits. Rockström (2010) believe that three of these boundaries have already been breached: climate change, the nitrogen cycle, and biodiversity loss.[2] There is presently 390 parts per million (ppm) of carbon dioxide in the atmosphere, with a limit of 350 ppm. Humanity has also past the planetary

[2]The other six processes with limits are: depletion of stratospheric ozone, land use change, freshwater use, ocean acidification, air pollution from aerosol loading, and chemical pollution.

boundary for a healthy nitrogen cycle, removing 121 million tonnes of nitrogen per year (largely used for fertilizer in food production), with a proposed boundary of 35 million tonnes per year. The rate of biodiversity loss is currently over 100 per million species per year, with a proposed boundary of 10 per million species per year. Put another way, there has been a loss of 52 per cent of mammals, birds, reptiles, amphibians and fish species between 1970 and 2010 (WWF 2014). Such statistics and examples emphasise the impact that human beings are collectively having on the planet.[3] These processes are operating in a historical and political context, with particularly important implications when it comes to continuing world population growth.

In the last 250 years the human population has risen seven-fold. When Malthus wrote his famous *Essay on the Principle of Population* (1798), the world population was under one billion people. By 1900 it had reached 1.7 billion people (UN 1999), by 2000 it had reached 6 billion, and it took just 12 years to increase from 6 to 7 billion. As of 2015 there are 7.325 billion people on the planet (UN 2015). The Population Division of the United Nations' Department of Economic and Social Affairs *2012 Revision* predicts a slowing down of the growth rate, such that humanity will reach 9.6 billion in 2050 and 10.9 billion by 2100 (UN 2013).[4]

Literature on stabilising population stresses the connections between stabilising population and social justice. For example, a more stable population is linked to increasing social stability, the reduction of child mortality and the alleviation of poverty. More stable populations are also linked to empowering women via gender equality, increasing access to education (particularly for females), improving maternal health and access to contraception. Furthermore stabilising population is linked to the development of green technologies that would enable the resources and energy needs of the global population to be met without disastrous implications for the Earth's ecosystems and climate (Shapiro 2012; de la Croix 2014; Oded 2011; Rosling 2010).

The WWF's Living Planet Report (2014) states that *current* human activity needs one and a half Earths to sustain it. That is, humanity is already using nature's gifts faster than they can be renewed. If low-income countries are able to be lifted out of poverty the "dual effect of a growing human population and high per capita Footprint will multiply the pressure we place on our ecological resources" (WWF 2014: 12). It is useful to recall Ehrlich/Holdren's (1974: 720) formula for calculating the environmental impact of humans on Earth: *Population* × *Consumption*

[3]Refer to UNEP's GEO5 report (2012) for more information and statistics.

[4]The *2012 Revision* states that the predicted stabilisation at 10.9 billion in 2100 is based on a "medium-variant projection" that assumes a "decline in fertility in many countries where large families are still prevalent" (UN 2013). This scenario has changed significantly since the *1998 Revision*, which predicted that world population would stabilise at 10 billion in 2200 (UN 1999). If the rate of one billion people every 12 years continues, the population will reach 10 billion at 2044, and 14 billion by 2100.

(or Affluence) × *Technology*.[5] Ultimately to reduce the impact of humans on the environment, the global community must decrease or stabilise population, decrease consumption, and/or improve technology such that 10 billion or more humans can satisfy at least their basic needs in non-harmful ways (Hart 2007: 31).

In this chapter the term 'global ecological crisis' is used to refer to the widespread destruction that humanity as a whole is causing to their environment. This section has posited that the impact that humanity has on their planet is tied to issues of social justice. It has suggested that addressing the global ecological crisis requires stabilising world population, addressing structural injustices in the world system, and developing ecologically harmonious ways of living. In order to move toward such solutions, one must consider the economic, political, historical and social factors behind present world systems. These dimensions will be elaborated in sections that follow.

3.3 Ecological Crisis as Structural Violence

Are people responsible for the ecological crisis, or are institutions? In this section, two theories will be synthesized to shed light on the causes of the global ecological crisis, viewed as a form of structural violence. It will consider the varying power of people and institutions to perpetuate or mitigate it, asserting that *both* people and institutions are responsible, via a complex of multilayered and multidirectional relationships.

Arguably, the global ecological crisis is a structural form of violence in the sense that "*no specific actors* are indicated, and … *no specific motivation* is necessary" (Galtung 1980: 183).[6] In the dominant neo-liberal capitalist system, normalized production and consumption habits of industrialized societies, supported by an international legal and economic framework, feedback into the system in ways that encourage the maximization of short-term profit for some individuals over the long-term health of the ecosystem (Chomsky 1999). In this form of capitalism one might blame the global ecological crisis on legal and economic structures, and on corporate and governmental institutions. Yet such structures and institutions are inseparable from the humans that accept them, operate within them, and who can work to change them. That is, within those corporations, governments and legal systems are people whose actions, while cultivated and operating within those structures, can also work to change them. Hence one might also blame individual

[5]This formula was first published in P. R. Ehrlich and J. P. Holdren (1974: 720). Originally "Consumption" is used rather than "Affluence": resource consumption = population × consumption per person; and hence: environmental impact = population × consumption per person × environmental impact per consumption. Affluence is used in later works by Ehrlich due to the handy acronym PAT (rather than PCT)—see also Ehrlich (1990: 58, 273).

[6]Emphasis is Galtung's. Galtung's 1980 article is a follow-up on his influential 1970 article "A Structural Theory of Imperialism." This chapter draws from both.

people and groups for the crisis, each who makes decisions in serving their own personal interests, in their roles as consumers, employees, shareholders, superannuation holders, CEOs and staff of corporations.

Theories of structural violence grapple with the 'emergent' properties of global systems, through which feedback mechanisms bring about unintended consequences. This dynamic relates to the work of complex systems theorists, who describe the way that parts can influence a whole, and a whole can influence the parts, with neither completely determined by the other (Thrift 1999). Out of interacting components emerges a property that "couldn't have predicted from what you know of the component parts," explains Chris Langton, that "feeds back to influence the behaviour … of the individuals that produced it" (cited in Thrift 1999: 33–34).[7] This multidirectional causation is useful in its application to the global ecological crisis explored as a form of structural violence. It illuminates tensions between individual short-term decisions and the collective long-term consequences of those decisions. This will be further interrogated in the analysis that follows.

For the purposes of this discussion the author has selected two theories that combine to provide a framework for exploring the global ecological crisis as a form of structural violence. The first theory is Galtung's (1971) "Structural Theory of Imperialism," which provides a historical context for the relationships between low- and high-income countries, relationships that arguably have a continuing influence on poverty, environmental destruction and climate change negotiations in the world today. Inspired by dependency theorists such as Raúl Prebisch while teaching in Chile (see Galtung 2014), and in line with Wallerstein (1974) and other world-systems theorists, Galtung's theory explores the dynamics of power between Centre and Periphery nations.[8] Galtung's theory was selected as the foundation for this model due to its extensive influence and due to Galtung's prominence in the field of peace and conflict studies. Insights from more recent theories that build on Galtung's framework will be integrated into the analyses below.

Building on the foundation laid by Galtung's theory, Kahn's (1966)'s "Tyranny of Small Decisions" was selected due to its explanatory value in terms of parts and wholes separated in time and space. It complements Galtung's theory in illuminating the bottom-up power within structural violence, providing further insights into the interlinking macro and micro dimensions of the global ecological crisis. The combination of theories will be used to identify those with power to help

[7]For example, out of the interactions of cells emerges an organ, which influences the behaviour of the cells. Out of the interaction of organs emerges a human body, which influences the behaviour of the organs. Out of the interaction of humans emerges a culture; which influences the behaviour of its humans.

[8]Theoretically Galtung's theory applies to any form of structural imperialistic power relations, including between two groups or two people, but he applies it primarily to the relationship between nation states. Galtung (1980: 184) notes that his theory "indicates *what* to look for if imperialism is at work, not *where* to look for it." Imperialism, here, might be in the sense of economic as well as the "political, military, communicative, cultural and social."

address the crisis. These theories will be considered separately, adding to them contemporary and complementary theories, and building a diagrammatic representation at each stage.

3.3.1 A Structural Theory of Imperialism

In his "A Structural Theory of Imperialism" (1971), Galtung divides nations into the *Periphery* (P) and the *Centre* (C), each having within them a periphery (p) and centre (c).[9] People who are in the *centre of the Centre* (cC) are posited as being a dominant power, the most influential people in the world. This might take the form of people who own large amounts of capital, corporate executives, governments, influential media persons, and academics—people with the power to influence and make decisions that affect the masses.[10] The people in the *periphery of the Centre* (pC) are the public majority of high-income countries—people who work, consume and live within the norms of the structure. The people in the *centre of the Periphery* (cP) are the more powerful people in low-income countries—people who benefit from selling the country's resources and labour to the wealthier nations. Finally, the people in the *periphery of the Periphery* (pP) are those with the least power, the four billion people at the bottom of the global pyramid of material wealth.

Galtung describes a Conveyor Belt pumping resources (human and natural) from the *periphery of the Periphery* (pP) to the *periphery of the Centre* (pC). This is indicated by the top arrow from P to C. Galtung (1971: 83) describes a *harmony of interests* between pC and the cC, and between the cP and cC—indicated by the unbroken lines in the figure. He also describes a *disharmony of interest* between the pP and cC, and between the C and P—indicated by the broken lines between C and P. Cash crops such as coffee, cocoa and cotton are examples of this conveyor belt in action. Cash crops benefit people in the pC, who can buy cheap coffee, chocolate and clothes, and the owners of those companies in the cC reap most of the profits. Meanwhile people in the cP benefit from the agreements, while people in the pP often have little choice but to work long hours in terrible conditions for a very low wage, or worse. *Structural Adjustment Programmes* (SAPs), often linked with neo-liberal agendas, might be considered a further example of this theory.[11]

Figure 3.1 builds on Galtung's (1971: 84) diagram to illustrate the dynamics and power relationships within Galtung's "Structural Theory of Imperialism".

---

[9]In some world systems theories such as Wallerstein (1974) use the word 'core' instead of 'centre', and include 'Semi-periphery' nations, representing expanding economics such as Brazil, Russia, India and China. The dynamics of the models still work in a similar fashion, and the addition of Semi-Periphery adds unnecessary complexity to this particular analysis.

[10]Leslie Sklair's conception of a powerful Transnational Corporate Class might be a useful way to conceive of the influential people in the centre of the Centre (this will be returned to in Sect. 3.2).

[11]Critics such as Pamela Sparr (1994) point out that non-industrial countries are growing food and produce goods for the industrialized countries, at the expense of their own people.

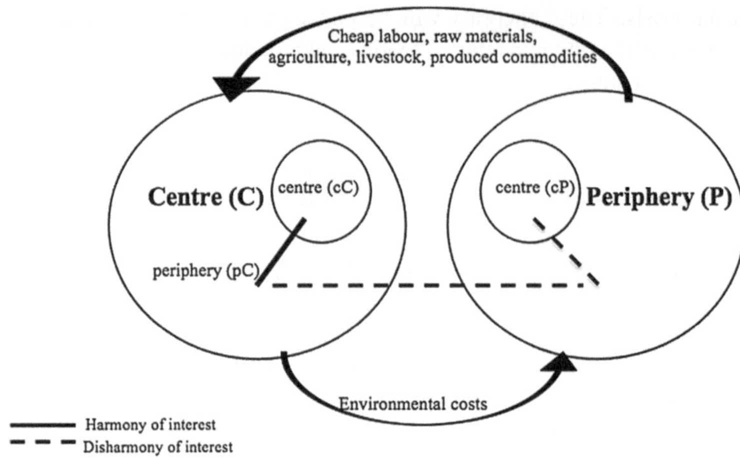

Fig. 3.1 Dynamics of Structural Imperialism. *Source* The author, adapted from Galtung (1971: 84)

Figure 3.1 has been adapted to recent research that shows 'environmental costs' flowing from the Centre to the Periphery—indicated by the lower arrow from C to P. Although Galtung's theory did not address environmental issues, more recent research such as the extensive work of Andrew K. Jorgenson and colleagues (e.g. Jorgenson 2006; Jorgenson/Clark 2011; Jorgenson/Givens 2014) builds on this and other world-systems and dependency theories. Jorgenson (2006: 687) posits a "structural theory of unequal ecological exchange" using the example of deforestation to make a case that "more-developed countries partially externalize their consumption-based environmental costs to less-developed countries which increase forms of environmental degradation within the latter" (704). The language of more- and less-developed countries is carefully selected. Jorgenson clarifies that this exchange, "partly a function of the historical legacies of colonialism," is not a "binary relationship".

Since Galtung wrote this paper middle-income countries like Brazil and China have risen in prominence on the world stage.[12] Instead of a dual separation between high and low income countries, one finds a continuum from high-income countries through high-middle, middle, low-middle and low-income countries. Jorgenson posits the uneven ecological exchange as taking place "cumulatively between relatively more-developed countries and less-developed countries" (Jorgensen 2006: 692). Jorgenson provides empirical evidence of a "core/periphery hierarchy". He observes that higher levels of organic water pollution, higher levels of infant mortality, lower levels of secondary education, correlate with higher percentages of export commodity concentration and higher levels of agricultural production. He

[12]These might be considered in terms of the 'semi-periphery' in Wallerstein's world systems model.

considers all these factors to be “largely a function of a country’s position in the core/periphery hierarchy” (Jorgenson 2004: 280). These factors feedback into perpetuating poverty, population increases and environmental destruction, for example due to the instability caused by infant mortality, the lower levels of education and higher levels of pollution. Clearly, if the ecological crisis is going to be mitigated then the feedback looks perpetuating injustices embedded of the world system of production and trade will have to be disrupted and new more equal relationships developed.

Another impact of structural inequality on the global ecological crisis can be seen in the failure of countries to establish an international agreement and framework for mitigating climate change. Parks/Roberts (2010) argue that the reason climate negotiations have largely failed is connected to the injustice embedded in the world economy, which “condition a state’s willingness and ability to participate in such arrangements.” Parks and Roberts observe that “a growing number of developing countries have called for a recognition of (and/or remuneration for) a so-called ‘ecological debt’ that the North owes the South” (142). If high-income countries have consumed fossil fuels in order to build infrastructures and housing, then why should low-income countries have to pay the costs? Parks and Roberts suggest that “climate change negotiations must be broadened to include a range of seemingly unrelated development issues such as trade, investment, debt, and intellectual property rights agreements” (134). Negotiating on global climate change is likely to require “wealthy industrialized states to shoulder a significant part of the cost of the transformation in developing countries” (Held/Hervey 2009: 2). A “hybrid proposal” suggested by Pew Center for Global Climate Change would be to implement this through a mixture of “responsibility based on past and present emissions, carbon intensity and countries’ ability to pay” (Parks/Roberts 2010: 152).

An everyday example will be useful for illustrating how these concepts tie together. Consumers in the periphery of the Centre may or may not know that a $5 cotton t-shirt is likely to involve sweatshop workers and cotton farmers working in near-slavery conditions, deforestation and the destruction of top soil, a significant amount of water consumption and carbon emissions from production through the transport from, for example, Brazil to China to Australia, and even its disposal. These connections are remote. The benefits are experienced by the people in Centre countries, for example in being able to purchase a low-priced t-shirt, work for a business involved in this production process, or receive profit from investments in companies involved. From a broader perspective one can draw distant yet cumulatively-influential connections between businesses and people benefiting from $5 t-shirts and ongoing social and ecological justice issues faced in Periphery countries. For example, the underpayment of workers in factories and on farms perpetuates poverty, and poverty as discussed earlier is linked to increasing populations. Furthermore, the production and consumption of the $5 t-shirt connects to ecological sustainability issues including climate change, for example as a result of the carbon based fuels to make and delivery the t-shirts, and the way that the unequal relationships between Periphery and Centre countries feeds into difficulties involved in climate change negotiations and action. Responsibility of the social and ecological injustice represented by the $5 t-shirt is dispersed among powerful economic and political institutions, companies and investors, governments, citizens and consumers. A single person is powerless against them.

While the decision made by one consumer to purchase a $5 t-shirt is of minute consequence, when multiplied by a million or 100 million people it can be the fuel that perpetuates sweatshops and destroys ecosystems. The consumer's decision may or may not be informed of these broader social and environmental costs. He or she may justify it through a neoliberal lens that considers it part of the Periphery's path to development. He or she may be aware of the impacts but be constrained by the financial burdens of high-income countries (such as mortgages, school fees). Or he or she may feel too small to make any real difference. Alternatively he or she may wish to make more ethical purchasing decisions but not know of any particularly ethical options available. This is an example of what Kahn called the "tyranny of small decisions."

3.3.2 *Tyranny of Small Decisions*

In 1966, Kahn postulated a gap in the dominant economic modelling of supply and demand, in its prioritization of short-term desires over long-term interests. Kahn (1966: 23) describes it as an "inherent characteristic of the market" that had not at that time been identified as, in some circumstances, producing a "defective or possibly objectionable allocational result." In exploring tensions between "private wants and public needs,"[13] Kahn points out that decisions which are smaller in size, scope and time, for example, an individual consumer's purchasing choices, can collectively have a larger result that impacts on the individual in ways that he or she would not choose if presented the choice as a whole.

Kahn uses an example with personal relevance to him, which is useful for explaining the theory. Kahn lives in Ithaca, a city in upstate New York. Until 1961 a railway operated that was the "one reliable means of getting into and out of Ithaca in all kinds of weather" (26). Due to individual customer decisions to save time or money by flying or driving, the train service was no longer financially sustainable and was shut down. Kahn explains that his own "introspective experiment" is proof that at least one customer (himself) would have been willing to pay extra (for example an annual fee), in order to keep the railway running. Each person's choice to take a flight or drive a car "had only a negligible effect on the continued availability" of the railway, and therefore it would have been "irrational … to consider this possible implication of his decision" (26). Kahn emphasizes the "necessity of looking at the process in broader terms than does the market, and possibly substituting 'large' for piecemeal accumulation of 'small' decisions" (25).

[13]Kahn notes that this is the title of readings edited by Edmund S. Phelps rev. ed., Norton, New York: 1965.

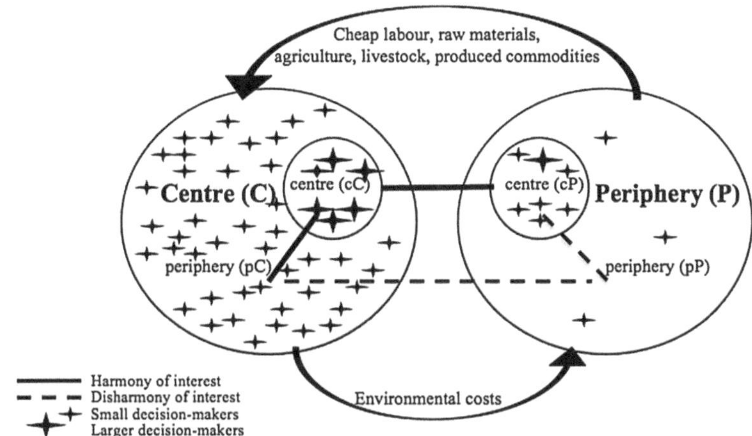

Fig. 3.2 The 'Tyranny of Small Decisions' maintaining injustice. *Source* The author, expanding Fig. 3.1

This broadening of decision-making can apply to individuals, for example consumers and CEOs, as well as decision-making by groups such as by governments of nation states.

Figure 3.2 builds on Fig. 3.1, using stars to represent the rough distribution of power that can be inferred by these theories.

In Fig. 3.2, the stars represent decision-makers, with larger stars representing decision-makers with a larger influence, and smaller stars represent decision-makers with a smaller influence.[14] The figure posits that the choices of people in the centre of the Centre have the most impact globally per decision, but a relatively small number of powerful people make these decisions. Decisions made by people in the periphery of Centre have less impact per decision but potentially have the most significant impact when joined together. People in the *centre of the Periphery* (cP) also have a large impact in perpetuating the oppression of the *periphery of the Periphery* (pP) for their own benefit. The pP have considerably less impact per decision, however they still have power (for example as exercised in the so-called Arab Spring).

The estimated distribution of power is supported by Sklair's (2002: 145) theorisation of a *transnational capitalist class* (TCC). Based on this theory, one could consider the large decision-makers in the Centre to be comprised of the four interlocking fractions: (a) "the corporate fraction"—shareholders and executives, people who own and control the major corporations; (b) "the state fraction"—"globalizing bureaucrats and politicians"; (c) "the technical fraction"—"globalizing professionals"—such as scientists, academics, and skilled

[14]Not larger in the sense of broader and longer term, but larger in terms of their impact on other people and on the planet.

workforce; and (d) "the consumerist fraction"—which includes the media, merchants and wealthy consumers. Through their money and influence these decision-makers have a greater power to maintain or change global structures than decision-makers with less money and influence.

An example of the power of this group is the scepticism toward *anthropogenic global warming* (AGW) that has developed "among laypeople and policy makers". This has been generated by a "loose coalition of industrial (especially fossil fuels) interests and conservative foundations and think tanks," assisted by "a small number of contrarian scientists", "conservative media and politicians" and "a bevy of skeptical bloggers" (Dunlap 2013: 692). Alternatively, if such a group was motivated to do so, they could use their influence to ignite enthusiasm for green technology, divestment from fossil fuels, and influence the political will for governments to implement policies to address socially and ecologically unjust structures.[15]

Environmentalist Tim Flannery observes that the abilities and costs of clean energy, such as wind and solar energy, are now comparable to coal and oil. In the way of its adoption are vested interests. Vested interests can be direct—for people such as CEOs' and shareholders' monetary rewards and dividends derived from businesses that profit from exploiting the planet (through fossil fuel industries, monocropping, deforestation for livestock farming, offshoring of wastes, etc.). Vested interests can also be indirect—for customers, civilians and governments for example allowing them to buy cheap oil and cheap food, via superannuation funds invested in these businesses, and via tax collected from the selling of such goods. For example, the Australian government and Australian people benefit from coal exports via tax the coal companies pay, via jobs the industry creates, etc., yet the impact of coal on the environment are long-terms costs that will be shared by all (see Pearse et al. 2013). Such direct and indirect interests are standing in the way of the personal and political will to invest in making changes in lifestyle, investment, policies, taxes etc., directed at addressing the global ecological crisis.

In respect to the global ecological crisis, the tyranny of small decisions can be observed as another factor standing in the way of polices to mitigate climate change. Held/Hervey (2009: 5) explain that it is "extremely difficult for governments to impose large-scale changes on an electorate whose votes they depend on, in order to tackle a problem whose impact will only be felt by future generations." They describe the issue as "short-termism," referring to the tendency for policy debates and implementation of policies to be limited by the short-term nature of political cycles. Politicians have to please their electorate in order to get voted back into power, incentivising governments to avoid implementing policies that may not be in the constituency's direct interests.

[15]This is not to say that people in the cC cannot ignite such enthusiasm themselves—one can certainly see the power of the people in grassroots movements advocating for these changes. The point is that the more influence a person has, the more difference that their everyday decisions can make.

Applying Kahn's tyranny of small decisions to environmental degradation, Odum (1982) makes a plea for less reductionist and more holistic approaches to research and decision-making. Odum considers the cumulative effects that the small decisions of individuals may have on the society or environment, reflecting briefly on the examples of air and water pollution, desertification, and management of fisheries. The phenomenon is far-reaching. The field of medicine has a tendency to focus on "single-cause and single-effect" with "modest emphasis on total body responses" (Odum 1982: 728). In academic research, Odum (729) points out that grants and tenure tend to be geared to projects that favour the short-term over the long-term, and specific outcomes over projects that impact on a broader level. This is understandable in terms of the 'small decisions', which benefit individuals in the short-term, however the result is a significant gap in the big picture.

The big picture is this: the unjust structures of the world system and the short-term orientation of politics and personal profit are obstacles to addressing the global ecological crisis. Galtung locates the unjust flow of human and natural resources at the level of nations, and sheds light on the historical context of these relations. The unjust flow of environmental exploitation stands in the way of climate change policies being successfully negotiated between countries. The unjust flows of natural resources and cheap labour from low- to high-income countries, perpetuates poverty in low-income countries and prevents their populations from stabilizing. This structural violence is maintained by small decisions made across many different levels and locations across global society. Kahn's identifies a gap in supply and demand economics that causes a separation between short-term and long-term motivation for decisions. It points to way that consumers, citizens, employees and investors everyday decisions feed into institutions and structures, skewing the supply-demand function of markets when it comes to longer-term outcomes. This interaction between small decisions and larger structures has led to the global ecological crisis. The final stage of the argument in Sect. 3.4 considers what insights this synthesis of theories may offer as to strategies for mitigating the ecological crisis and moving towards ecological peace.

3.4 Re-orienting Decisions Towards Ecological Peace

Section 3.3 examined the dynamic of multi-directional and multi-levelled interactions that are causing the global ecological crisis. This section explores how such interactions might be re-oriented towards a vision of ecological peace, in light of the insights offered by the above synthesis of theories. It will take an imaginative leap, considering a deep cultural shift toward holism that might help to motivate the re-orienting of decision-making proposed.

With the analysis from previous sections in mind, Figs. 3.3 and 3.4 explore the model from Sect. 3.3 with two snapshots in time: short-term and long-term.

Figure 3.3 posits that in the short-term the world system is very good for the *centre of the Centre* (cC), good for the *periphery of the Centre* (pC), and pretty

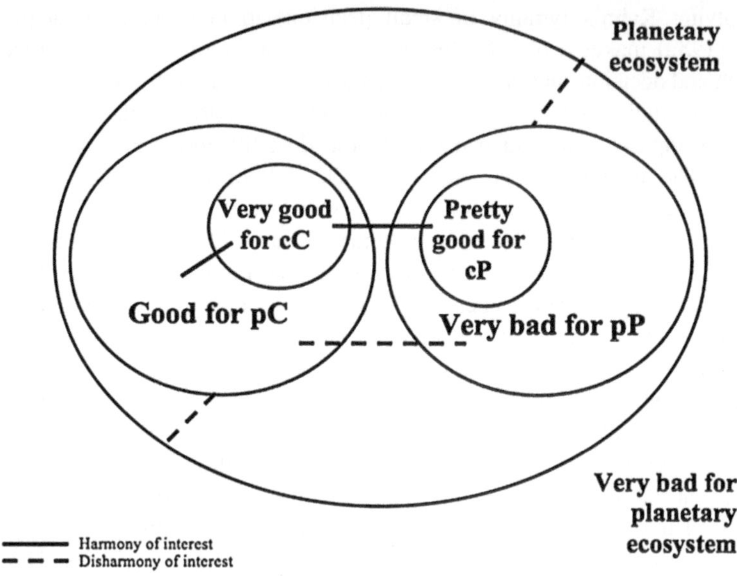

Fig. 3.3 Beneficiaries and benefactors in the short-term. *Source* The author

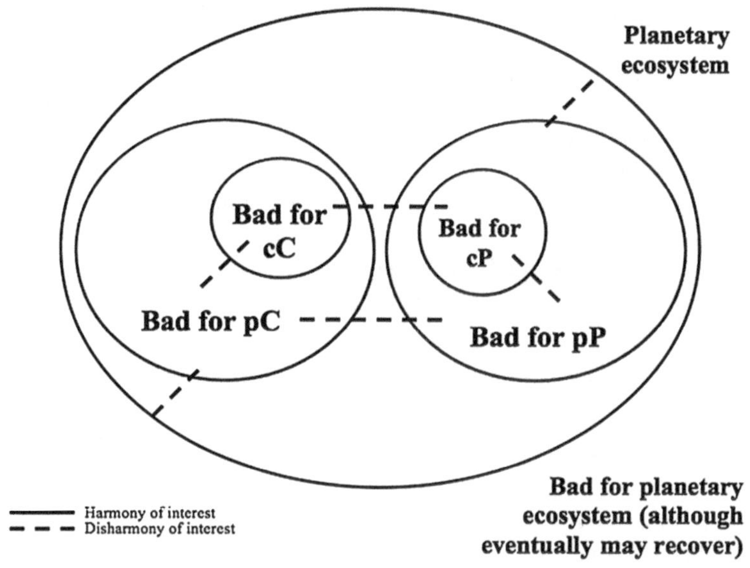

Fig. 3.4 Beneficiaries and benefactors in the long-term. *Source* The author

good for the *centre of the Periphery* (cP). For these groups, increases in production, increases in consumption and increases in profit, are good. This world system is also very bad for people in the periphery of the Periphery (pP), who are exploited and live in poverty. Finally, this status quo is destroying the planetary ecosystem. Figure 3.4 posits that in the long-term, this world system is not in the interests of any parties, whose habitat and resources will be destroyed. These figures point out that it is in the interests of *all people*, including the world's most powerful (assuming they care about anyone or anything beyond themselves), that the world system evolves into one that is more sustainable and hence is also more just.

What can be done to change the world system? A common thread can be seen in the recommendations of Galtung, Kahn and peace scholars: a tendency toward holism. Galtung (1971: 88) suggests solutions to situations of structural imperialism lie in the "social totality." He emphasizes the need to explore "the totality of the effects of an interaction process" including the economic, political, military, educational and communication dimensions, as well as the cultural, social and psychological effects. Echoing Galtung and building on Kahn, Odum suggests that the key to avoiding the problem of small decisions lies in developing a holistic understanding of the context and consequences of decisions. Peace educator Reardon (1988: 60) sums up this view: "We must learn to see ourselves as a part of, not apart from, our planet and all of its inhabitants." In sum, bridging the micro and the macro, the local and the global, short-term and long-term, decision-makers must come to see how their small decisions accumulate to bring about global and long-term outcomes for Earth and all living beings.

This is an important pattern worthy of deeper consideration. The pattern is a connection between parts and wholes in time (short-term and long-term) and in space (personal and global). This pattern as applied in process philosophy, panentheistic theology, deep ecology and macro history (Clayton/Peacocke 2004), challenges the core metaphysical assumptions on which the current world system is based. Instead of assuming the self is an individual separate being, acting in its own self-interest, the pattern contextualises the self in a community of changing relationships. This view points out that, in the long-term, personal self-interest is also that which is in the best interests of the global community. In this view, the temporal self as experienced in bodily form is just one expression of the infinite Self. Such an understanding can be derived from the simple observation that one is inseparable—in time and space—from the rest of the universe. It leads to the understanding of an *ecological self*, inseparable from the ecological systems throughout which it cannot exist. It also leads to the understanding of a *cosmological self*, part of a unified story of expanding consciousness and creative evolution. This view is found in a growing body of scholarship (Swimme/Berry 1992; Tucker/Grim 1994; Birch/Cobb 1981; Daly/Cobb 1989) that suggests that a process paradigm, ecological worldview or a 'new story' can help to address the global ecological crisis.

What kind of world system might the efforts of more holistic decision-makers work to create? What would a just and sustainable alternative look like?

Figure 3.5 takes inspiration from Boulding's (1988) workshop on "Imaging a World Without Weapons", that considers positive images of the future to be like a magnet, attracting behaviour that toward the vision. In this model the author suspends thoughts on what should be considered 'realistic', and posits a system in

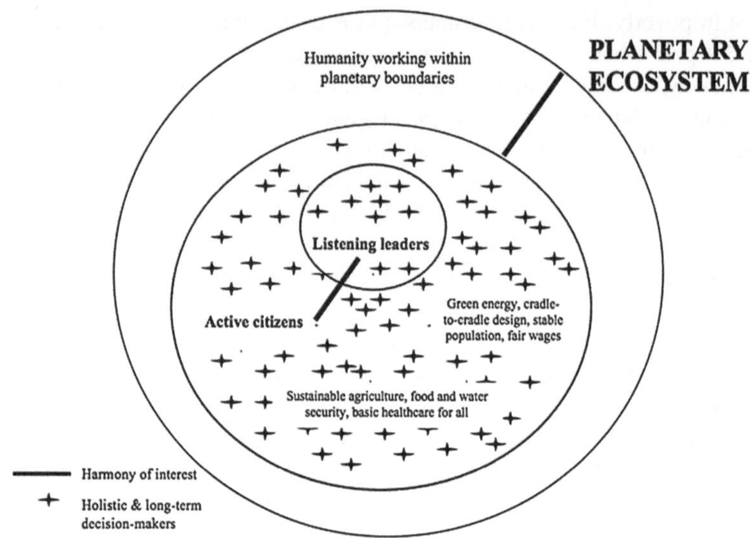

Fig. 3.5 An imagined just and sustainable world system. *Source* The author

which there would no longer be Centre and Periphery nations operating with unequal exchanges.

In this model people and institutions across the world would interact in ways that are in harmony with the planet's ecosystems. It imagines leaders who listen, and citizens who actively participate in civil society. In this model there is a harmony of interest between the listening leaders and active citizens, and a harmony of interests between humanity and the planetary ecosystem. It imagines the use of sustainable and localised agricultural techniques, food and water security, and education and healthcare for all people. It imagines green energy solutions replacing fossil fuels, and ecological designers creating ways that humans can live within planetary boundaries and in ways that Earth is better off for it (e.g. see Cowan/Van Der Ryn 1996; McDonough/Braungart 2002). These would be shared across the world without patents, in the name of equality, creativity and acting in the interests of the whole. As such the imagined model would not create new dependencies. Such a vision would see individuals acting mindfully, ethically and with empathy in all their interactions (e.g. see Kaza 2009; Rifkin 2009).

Placing all of the nations (whether currently in Periphery or Centre) in one circle does not infer a homogeneous society, but an equal one. An appreciation for the diversity of lifestyles and cultures, of different ways of being in the world, is an important part of process-based worldviews (Griffin 1994). All nations would interact within a level playing field, based on the principles of social justice, universal human rights and an appreciation for the intrinsic value of all living things. By seeing the other aspects of one's Self, it is possible to be motivated to act in the broader interests of the ecological whole.

The optimism espoused by the model stands in stark contrast to the assumptions of dominant rationalist economic and neoliberal political models. Those with the most power to change the system are those most benefiting from the status quo (in the short-term). Why would anyone help to change structures if it is not in their immediate personal interests? The author does not have space to consider this question in detail here. The fact that many scientists, politicians, activists and consumers are already directing their research, policy decisions, advocacy and purchasing dollars toward the aim of global justice and ecological sustainability, indicates that while it may seem a high ideal, ecological decision-making is not impossible to achieve.

What kinds of changes might this involve? A myriad of literature on social and ecological justice offers plenty of examples. It is worth mentioning a few.

First and foremost, in the centre of the Centre. Politicians and policy makers could focus on long-term outcomes and put into action national and international agreements on issues considered above. For example, policies aimed at slowing deforestation to levels that match reforestation, subsidising clean energy alternatives, taxing carbon emissions, a cap and trade scheme, etc. (e.g. see Held/Hervey 2009). Governments could place limitations on the concentration of wealth and power in the hands of few. For example, by cracking down on tax havens, putting a limit on the size of corporations, enforcing international minimum and maximum wages, and ensuring that lobby groups and media are not interrupting democratic process (e.g. see Brand 2014). CEOs could invest in green engineering of their production and distribution processes, ensuring that all people are paid fairly and the planet is not exploited in any related processes (e.g. see Hart 2007). Academics might collaborate on interdisciplinary projects aimed at practical outcomes in mitigating the ecological crisis.

In the periphery of the Centre, citizens could promote such policy priorities by being active in expressing the care for the interests of future generations, even where it requires small personal sacrifices. As consumers they could take into account the social and environmental ethics when making purchases. As employees they could choose only to work for companies that are socially and ecologically just. In the centre of Periphery nations, individuals could crack down on corruption and enact laws and regulations to prevent the exploitation of the environment or of people. They could insist that natural resources such as rainforests, which provide ecological services to all of humanity, are maintained and paid for via contributions in higher-income countries (Held/Hervey 2009: 15). Poverty in the periphery of the Periphery could be addressed through education, health care, contraception, a fairer sharing of global resources, and the repayment of ecological debt by higher-income countries. People in the periphery of Periphery nations could then consciously choose to stabilise global population by choosing to have less children. An average of two children per family might become a global norm, a choice made by individuals in the interest of the global whole.

All of the above transformations could be achieved by spiral-upward process of (1) broadening the scope of individual decision-making to take into account of the interests of the whole; (2) individuals working to develop mediating structures to represent their collective political will toward common good and to help coordinate specific awareness and action campaigns for specific institutional, legal and cultural change; (3) influencing the reform of political, legal and economic institutions in ways that will feedback into step (1) in cultivating further decision-making and political will aimed at developing a more just and sustainable world system. Each step of such a transformation calls for a combination of the intertwining personal and political will toward long-term change. People across all sectors could encourage each other to put living beings and the planet before short-term profits, to celebrate altruism and shame people who have exploited other people or the planet in the name of personal wealth. A source of hope for such a holistic re-orientation of decision-making can be found in process scholarship crossing a broad range of disciplines, and the related intellectual movement that is attempting to reimagine and reinvent education, culture, society, art, health, philosophy, theology, psychology and nature.[16]

There is no space to explore, analyse, compare and evaluate the many efforts across the world working toward ecological peace. Suffice to say that the above sampling of ideas reflects some possible creative steps aimed at that direction. The point to be made is that if humans have the motivation to do so they can confront destructive social, political and economic institutions and evolve the world system to be more peaceful, socially just and ecologically sustainable. Strategies for motivating and implementing change toward ecological peace and developing integrated economic and political models to support it, are rich and exciting areas for further research and activism.

3.5 Conclusion

This chapter has explored two important aspects of the global ecological crisis: What is causing the crisis? Who has the power to solve it? Analysing theories and examples of structural violence and small decisions has pointed to the collective power of people to maintain or change institutions, industries, cultures and everyday actions that are causing the global ecological crisis. Galtung's "Structural Theory of Imperialism" provided a historical framework through which to understand the connections between issues of social and ecological justice and the world system. Kahn's "Tyranny of Small Decisions" shed light on the dynamics of everyday decisions that maintain those structures. The synthesis of theories pointed

[16]Process thinkers were addressing these questions at the "Seizing an Alternative: Toward an Ecological Civilization" conference in June 2015, hosted by the Center for Process Studies, in Claremont CA. This conference brought together Bill McKibben (creator of 350.org) with Vandana Shiva, Mary Tucker Evans, Herman Daly and the world's leading process thinkers including John Cobb Jr., David Ray Griffin, Catherine Keller, Phillip Clayton and Arran Gare. See conference program, at: https://www.ctr4process.org/whitehead2015/wp-content/uploads/2015/05/WH2015_online-program.pdf (26 September 2015).

out that in order for these solutions to be implemented, connections between the short-term and long-term, between the personal and global, must be made. Suggestions that such a shift might be motivated by a change in worldview or a 'new story' were very briefly considered along with some well-known examples of the types of decision-making that are likely to help mitigate the global ecological crisis. To sum up: addressing the crisis calls for governments, corporations and civilians to put global needs before personal interests, and to evolve structures in the interests of all. What the world system might look like through this shift in paradigm has been imagined. Unpacking positive visions of the future in greater detail, and experimenting with the ways this shift might come about, are questions for future papers and further research.

References

Birch, Charles; Cobb Jnr., John B., 1981: *The Liberation of Life: From the Cell to the Community* (Cambridge: Cambridge University Press).

Boulding, Elise, 1988: "Image and Action in Peace-Building", in: *Journal of Social Issues*, 44, 2: 17–37.

Brand, Russell, 2014: *Revolution* (London: Random House).

Carson, Rachel, 1965: *Silent Spring* (Harmondsworth: Penguin Books).

Chomsky, Noam, 1999: *Profit over People: Neoliberalism & Global Order* (New York: Seven Stories Press).

Clayton, Philip; Peacocke, Arthur (Eds.), 2004: *In Whom We Live and Move and Have Our Being: Panentheistic Reflections on God's Presence in a Scientific World* (Cambridge: Wm. B. Eerdmans).

Daly, Herman E.; Cobb Jnr., John B.,1989: *For the Common Good: Redirecting the Economy toward Community, the Environment, and a Sustainable Future* (Massachusetts: Beacon Press).

de la Croix, David, 2014: "Fertility, Education, Growth, and Sustainability", in: Salvadori, Neri (Ed.): *The Cicse Lectures in Growth and Development* (Cambridge: Cambridge University Press).

Dunlap, Riley E., 2013: "Climate Change Skepticism and Denial: An Introduction", in: *American Behavioral Scientist*, 57,6: 691–698.

Ehrlich, Paul R.; Holdren, John P., 1974: "Impact of Population Growth", in: *Science*, 171: 1212–17.

Ehrlich, Paul R.; Anne H. Ehrlich, 1990: *The Population Explosion* (New York: Simon & Schuster).

Escobar, Arturo, 1997: "Cultural Politics and Biological Diversity: State, Capital and Social Movements in the Pacific Coast of Colombia", in: Lloyd, D.; Lowe, Lisa (Eds.): *The Politics of Culture in the Shadow of Capital* (Durham, NC: Duke University Press): 201–26.

Galtung, Johan, 1969: "Violence, Peace and Peace Research", in: *Journal of Peace Research*, 6,3: 167–91.

Galtung, Johan, 1971: "A Structural Theory of Imperialism", in: *Journal of Peace Research*, 8,2: 81–117.

Galtung, Johan, 1980: "A Structural Theory of Imperialism: Ten Years Later", in: *Millennium: Journal of International Studies*, 9,3: 181–96.

Galtung, Johan, 2014: "The Group of 77 at Fifty Congratulations", in: *UN Chronicle*, 51,1: 14.

Goldsmith, Edward; Allen, Robert, 1972: *The Economist's Blueprint for Survival* (Boston: Houghton Mifflin).
Gore, Al, 2006 (ed. by Davis Guggenheim): *An Inconvenient Truth: A Global Warning* (Los Angeles: Participant Media).
Griffin, David Ray, 1994: "Whitehead's Deeply Ecological Worldview", in: Tucker, Mary Evelyn; Grim, John: *Worldviews and Ecology: Religion, Philosophy, and the Environment* (New York: Orbis Books): 190–206.
Hart, Stuart L., 2007: *Capitalism at the Crossroads: Aligning Business, Earth, and Humanity*. 2nd ed (Upper Saddle River, NJ: Wharton School Publishing).
Held, David; Hervey, Angus Fane, 2009: *Democracy, Climate Change and Global Governance: Democratic Agency and the Policy Menu Ahead* (London: Policy Network).
Higgins, Polly; Short; Damien; South, Nigel, 2013: "Protecting the Planet: A Proposal for a Law of Ecocide", in: *Crime, Law and Social Change*, 59,3: 251–266.
Jorgenson, Andrew K., 2004: "Global Inequality, Water Pollution, and Infant Mortality", in: *Social Science Journal*, 41,2: 279.
Jorgenson, Andrew K., 2006: "Unequal Ecological Exchange and Environmental Degradation: A Theoretical Proposition and Cross-National Study of Deforestation, 1990–2000", in: *Rural Sociology*, 71,4: 685–712.
Jorgenson, Andrew K.; Clark, Brett, 2011: "Societies Consuming Nature: A Panel Study of the Ecological Footprints of Nations, 1960–2003", in: *Social Science Research*, 40,1: 226–244.
Jorgenson, Andrew K.; Givens, Jennifer E., 2014: "Economic Globalization and Environmental Concern: A Multilevel Analysis of Individuals within 37 Nations", in: *Environment and Behavior*, 46,7: 848–871.
Kahn, Alfred E., 1966: "The Tyranny of Small Decisions: Market Failures, Imperfections, and the Limits of Economics", in: *Kyklos*, 19,1: 23–47.
Kaza, Stephanie, 2009: *Mindfully Green: A Personal and Spiritual Guide to Whole Earth Thinking* (Lane Cove, N.S.W.: Finch Publishing).
Leopold, Aldo, 1949: *A Sand County Almanac, and Sketches Here and There* (New York: Oxford University Press).
Malthus, Thomas, 1798: *An Essay on the Principle of Population* (London: J. Johnson).
McDonagh, Sean, 1986: *To Care for the Earth: A Call to a New Theology* (London: Cassell Publishers Ltd).
McDonough, William; Braungart, Michael, 2002: *Cradle to Cradle : Remaking the Way We Make Things* (New York: North Point Press).
Meadows, Donella H.; Meadows, Dennis L.; Randers, Jorgen; Behrens III, William W., 1972: *The Limits to Growth; a Report for the Club of Rome's Project on the Predicament of Mankind* (New York: Universe Books).
Oded, Galor, 2011: *Unified Growth Theory* (Princeton: Princeton University Press).
Odum, William E., 1982: "Environmental Degradation and the Tyranny of Small Decisions", in: *BioScience, Issues in Biology Education*, 32,9: 728–9.
Parks, Bradley C.; Roberts, J. Timmons, 2010: "Climate Change, Social Theory and Justice", in: *Theory, Culture & Society*, 27,2–3: 134–166.
Pearse, Guy; McKnight, David; Burton, Bob, 2013: *Big Coal: Australia's Dirtiest Habit* (Sydney: New South Publishing).
Phelps, Edmund S., 1965 [Rev. ed.]: *Private Wants and Public Needs; Issues Surrounding the Size and Scope of Government Expenditure* (New York: Norton).
Programme), UNEP (United Nations Environmental, 2012: *Geo5 Global Environmental Outlook: Environment for the Future We Want* (Nairobi, Kenya: United Nations Environment Programme).
Rajagopalan, R., 2011 [2nd ed.]: *Environmental Studies: From Crisis to Cure* (New Delhi: Oxford University Press).
Reardon, Betty A., 1988: *Comprehensive Peace Education: Educating for Global Responsibility* (New York: Teachers College).

Rifkin, Jeremy, 2009: *The Empathic Civilization: The Race to Global Consciousness in a World in Crisis* (New York: Tarcher/Penguin).

Rockstrom, Johan, 2010: "Planetary Boundaries", in: *New Perspectives Quarterly*, 27,1: 72–74.

Rockstrom, Johan; Steffen, W.; Noone, K.; Persson, A.; Chapin, III, F. S.; Lambin, E.; Lenton, T. M.; Scheffer, M.; Folke, C.; Schellnhuber, H.; De Wit Nykvist, C. A.; Hughes, T.; van der Leeuw, S.; Rodhe, H.; Sorlin, S.; Snyder, P. K.; Costanza, R.; Svedin, U.; Falkenmark, M. L.; Corell Karlberg, R. W.; Fabry, V. J.; Hansen, J.; Walker, B.; Liverman, D.; Richardson, K.; Crutzen, P.; Foley, J., 2009: "Planetary Boundaries: Exploring the Safe Operating Space for Humanity", in: *Ecology and Society*, 14,2: 32 [online].

Rosling, Hans, 2010: "Global Population Growth, Box by Box"; at: TED@Cannes.

Shapiro, David, 2012: "Women's Education and Fertility Transition in Sub-Saharan Africa", in: *Vienna Yearbook of Population Research*, 10: 9–30.

Sklair, Leslie, 2002: "Democracy and the Transnational Capitalist Class", in: *Annals of the American Academy of Political and Social Science*, 581,1: 144–157.

Sparr, Pamela (Ed.), 1994: *Mortgaging Women's Lives: Feminist Critiques of Structural Adjustment* (New Jersey: Zed Books).

Swimme, Brian; Berry, Thomas, 1992: *The Universe Story: From the Primordial Flaring Forth to the Ecozoic Era—a Celebration of the Unfolding of the Cosmos* (New York: Harper Collins).

Thrift, Nigel, 1999: "The Place of Complexity", in: *Theory Culture Society*, 16,31: 31–69.

Tucker, Mary Evelyn; Grim, John A. (Eds.), 1994: *Worldviews and Ecology: Religion, Philosophy, and the Environment, Ecology and Justice Series* (New York: Orbis Books).

UN, 1999: *The World at Six Billion* (New York: United Nations, Department of Economic and Social Affairs, Population Division).

UN, 2013: *World Population Prospects: The 2012 Revision* (New York: United Nations, Department of Economic and Social Affairs, Population Division).

UN, 2015: *Population, Consumption and the Environment, Wall Chart* (New York: United Nations).

Van der Ryn, Sim; Stuart Cowan, 1996: *Ecological Design* (Washington, D.C.: Island Press).

Wallerstein, I, 1974: "The Rise and Future Demise of the World-Capitalist System", in: *Comparative Studies in Society and History*, 16,4: 387–414.

WWF, 2014: *Living Planet Report 2014: Species and Spaces, People and Places* (Gland, Switzerland: WWF International).

Chapter 4
Loving Nature: The Emotional Dimensions of Ecological Peacebuilding

Katharina Bitzker

Abstract While the mainstream environmental discourse seems to have taken a technocratic turn during recent years and promotes shallow ecological solutions which often fail to address underlying emotions as drivers for structural and cultural violence, there is also plenty of evidence of emerging research that offers a broader, more holistic perspective and puts the emotional/affective component—some call it love of nature—at its centre. This essay explores how the experience of 'loving nature' has been conceptualized in some of the literature pertaining to cultural ecology so far and how these experiences translate (or do not translate) into different daily practices that are conducive to ecological peacebuilding and ultimately a 'happy planet'. Drawing on the work of anthropologist Kay Milton, one of the core questions becomes: is it a mere coincidence who is actively engaged and concerned with the well-being of nature and who might be more or less indifferent to the current ecological degradation? Loving (or at least respecting) nature and acting accordingly appears to be a prerequisite for love between humans at this point in time. The current global ecological degradation reminds us that focusing on human-human aspects of love alone tends to neglect the simple fact that we are destroying what gives us life—while being proud of our loving behavior towards other human beings. This essay highlights why it might be important to broaden current anthropocentric models of love and shift to an ecological model of loving, how practices of resistance and complicity are embedded in an emotional field, why some sort of value coordinate system for 'sustainable loves' might be needed in the global north and the importance of embodiment/embodied emotions for our capacity to experience love or feel cut off from love.

Keywords Loving nature · Ecological view of emotions · Embodiment · Practices of resistance · Complicity · Biophilia · Happy planet index · Sustainable loves

---

Dr. Katharina Bitzker, Mauro Centre for Peace and Justice, University of Manitoba, Canada; Email: katharina.bitzker@hotmail.de. I would like to thank Prof. Derek Johnson, Dept. of Anthropology at the University of Manitoba, for the many helpful comments and discussions regarding this essay. Moreover, I thank Terry Mitchell for always keeping a protective eye on my stream-like writing style.

H.G. Brauch et al. (eds.), *Addressing Global Environmental Challenges from a Peace Ecology Perspective*, The Anthropocene: Politik—Economics—Society—Science 4, DOI 10.1007/978-3-319-30990-3_4

4.1 'Speaking Our Own Truth Is Like Oxygen'

The room is tightly packed with nearly 100 people. They form a circle around four objects placed in four different quadrants. A stone, dry leaves, a stick and an empty bowl. Although I have been participating in this ritual called 'The Truth Mandala' so many times, even facilitating it myself in different settings, I am always deeply moved and have cried every single time. But it looks like as if this is going to be my new personal sobbing record. The renowned Deep Ecologist Joanna Macy (1998, 2007) has come to Germany to teach 'the work that reconnects', as she calls it, and this weeklong conference has attracted people from all walks of life who share a deep concern for nature. Joanna Macy picks up the objects and explains what they stand for. The stone stands for fear because this is how our hearts feel when we are afraid: tight, hard, cold. The dry leaves represent our sorrow and mourning for all we are losing and have lost. There is great sadness within us for what we see happening to our world. The stick symbolizes our anger and outrage which need to be spoken for clarity of mind. And the empty bowl is a reminder of our own hopelessness, emptiness. In the middle of the circle a cushion is placed—an invitation to voice sentiments that are important but are not represented through the four objects. "You may wonder where hope is", she asks with a smile. "The very ground of this mandala is hope. If we didn't have hope, we wouldn't be here." We take turns stepping into the circle, holding the objects in our hands, breathing, gathering ourselves and then often falling apart in the most graceful way. It is amazing to witness all these people expressing their pain for the world. It seems most global north citizens have become so used to people caring only about their own little concerns, their own little world, that it moves everybody beyond words to finally understand—I am not alone in my concern for the world, I am not as freakish as I thought. For the next one and half hours we listen to a stream of expressions of fear, sadness, anger, hopelessness, and also little songs, poems, proverbs that people share as they sit on the cushion in the middle. "I am sad about the greediness I see all around me." "I am angry at this stupid government for blocking and reversing so many great ideas on how to protect nature." "I am afraid of the violence I see going on in my community." "I am sad that my best friend has died so young, without him I feel lost, he was my biggest ally." "Everyone around me thinks I am so optimistic but the truth is I feel so much despair most of the time when I look at what's going on in the world." "I feel everything that I am doing for the planet is not enough, it's a joke, I am lying to myself."

Whenever the person in the middle of the circle has finished, the rest of the group simply states, "We hear you." This creates a very different space of really 'being with'—not some sort of agreement or approval of what has been said (in terms of content) is required, no intellectual dissection starting with 'yes, but…', no pity for how bad the person has got it, just a plain 'we hear you'.

As the Truth Mandala draws to its end, Joanna Macy steps into the circle again. She picks up the different objects and says: "Please be aware of what you have been expressing and hearing. In hearing fear, you also heard the trust it takes to speak it.

In hearing sadness we heard in equal measure love because we only mourn what we deeply care for. In hearing anger we also heard passion for justice. And even the empty bowl, our own emptiness is to be honored because it means space for new ideas." She concludes the ritual with a beautiful take-home-message: "telling our own truth is like oxygen, it enlivens us. Without it we grow numb and confused" (Macy/Brown 1998: 103).

Had you encountered all these people during a more conventional conference, a more formal setting where such display of emotions would often be frowned upon, you would have never guessed the deep emotional undercurrents of their work. Probably you would have been drawn to their magnetic presence, acknowledged their tireless work for creating peaceful and ecologically sustainable societies, their enthusiasm, activism…And looking at myself and everyone else around me I could not help wondering—why do we so often hide these emotions that seem to be one of the major drivers of our work, our engagement? Why does it sound weird or make me feel awkward and vulnerable to write or talk about loving nature in an academic setting, whereas probably no one would bat an eyelid about a title such as 'the affective dimension of natural resource management'?

4.2 Introductory Remarks

While the mainstream environmental discourse seems to have taken a technocratic turn during recent years by promoting shallow ecological solutions which have led into a *cul-de-sac* (see e.g. the critical voices of postdevelopment thinkers like Sachs 2010 and Escobar 2011), there is also plenty of evidence of emerging research that offers a broader, more holistic perspective and puts the emotional/affective component—some call it love of nature—at its centre (see e.g. Drengson/Inoue 1995; Milton 2002; Nicholsen 2002; Macy 2007; Wegmann 2012). These voices are gaining more prominence as we currently seem to be in a time period where love as a creative and transformative power is making a comeback (or in some cases a first appearance) in many academic disciplines (see for e.g. Odent 2001; Maturana/Verden-Zöller 2008; Jónasdóttir/Ferguson 2014). The French researcher Odent (2001) has called this process the "scientification of love" and draws on anthropological evidence to examine reasons why some societies have been able to create loving bonds beyond the narrow anthropocentric (romantic) notion of love which is dominating much of our mental spaces right now. Moreover, many concepts such as sacredness, beauty, awe, relational ontologies, biophilia (see e.g. Bateson 1987; Kellert/Wilson 1993; Charlton 2008; Hinds/Sparks 2008) could be viewed as being rooted in the experience of loving nature.

For this essay I would like to explore how the experience of loving nature has been conceptualized in some of the literature pertaining to cultural ecology so far while spotlighting how these experiences translate (or do not translate) into different daily practices that are conducive to a 'happy planet' (Anielski 2007; Abdallah et al. 2012). My main objective is to sketch out a landscape regarding loving nature rather than focusing on one small area and providing an in-depth analysis of a particular issue. My starting point is the affluent global north, my home territory, and certain observations I have made there. Perhaps it might be more accurate to refer to shifting demarcation lines and what some authors (e.g. Sachs 2010) describe as the formation of the 'transnational consumer class'. As Sachs (2010: viii) vividly describes the lines that mark who belongs to the "global north/developed world/affluent" are transcending the borders of nation states.

> It comes as no surprise that the age of globalization has produced a transnational class of winners. Though they exist in different densities around the globe, this class is to be found in every country. In the large cities of the South, glittering office towers, shopping malls filled with luxury brands, gated communities with villas and manicured gardens, not to speak of the stream of limousines on highways or the never-ending string of brand advertisements, signal the presence of high purchasing power. Roughly speaking, half of the transnational consumer class resides in the South, and half in the North. It comprises social groups which, despite their different skin colour, are less and less country-specific and tend to resemble one another more and more in their behaviour and lifestyle.

The 'home perspective' I am adopting for this essay should not be read as an invitation to view global south or indigenous perspectives as being automatically more ecologically sustainable in some sort of 'noble savage' approach. The question, as so often, is rather how to interrogate privilege without re-centring it (Pease 2012). And although I cannot provide a full answer to this, I do believe that many peacebuilders who belong to the transnational consumer class might analyse their unearned privileges but shy away from actually living a more ecologically sustainable life. Or perhaps prefer to talk about sustainable development in the so-called global south.

I deliberately take a very broad definition of ecological peacebuilding. Personally, I find many initiatives—for example, the whole Transition Town movement (Hopkins 2008, 2011)—much more inspiring with regards to ecological peacebuilding than many of the officially labelled environmental peacebuilding initiatives. Many nature lovers would never dream of calling their projects 'peacebuilding' although they could probably easily compete with many projects on—what Denskus (2007) rather sarcastically refers to as—"the peacebuilding catwalks." It seems the previously criticized 'add gender and stir'—approach has been replaced by the new buzzwords of sustainable development/sustainable peace—now it is 'add environment and stir' within peace and conflict studies.

Through this essay I would like to highlight why it might be important to broaden current anthropocentric models of love and shift to an ecological model of loving, how practices of resistance and complicity are embedded in an emotional field, why some sort of value coordinate system for 'sustainable loves' might be needed in the global north and the importance of embodiment/embodied emotions for our capacity to experience love or feel cut off from love.

4.3 The Minefields of Separating the Inseparable and Describing the Ineffable

Pyle (2003: 213), after spending many pages on writing about reconnecting people and nature, concludes his essay with the following words: "ultimately, reconnecting people with nature is a nonsense phrase, for people and nature are not different things, and cannot be taken apart. The problem is, we haven't yet figured that out." This sentiment is echoed by many anthropologists, for example, Berglund (2006: 99) who points out that "it is impossible to establish the boundary between nature and culture empirically" and, like many other researchers in the area of cultural ecology, muses what we actually gain by clinging to this deeply engrained dichotomy. When I talk about 'loving nature' it might appear I ultimately imply that nature is an 'other', something separate, different from human beings. So I find myself stuck because what I would like to convey is actually quite the opposite: that our existences are so intertwined and create what the anthropologist Ingold (2011) describes as meshwork that I find myself amazed that a majority of people in affluent societies seem completely oblivious to this interconnectedness. And yet, not mentioning the nature/culture divide does not seem like a solution either—or as Pálsson (2006: 91) states: "dualisms don't disappear just because people stop talking about them."

Furthermore, there is a certain speechlessness when it to comes to writing about loving nature. Nicholsen (2002: 19) points out that "our relation to the natural world is in some important way nonverbal and unspoken. (…) Does this mean that to speak about that encounter is to objectify it rather than express our experience directly?" In fact, many nature lovers will agree with Adorno's (1997: 69) observation that "the 'How beautiful!' at the sight of a landscape insults its silence and reduces its beauty."

Writing about love in general is not exactly easy. Like many phenomena we human beings feel drawn to and take great delight in (humor comes to mind) it seems the magic is sucked right out of the phenomenon by the time there is an attempt to put the process into neatly packed, scientifically labeled boxes. What is love anyway? It would exceed the scope of this essay to discuss the different definitions of love—or rather attempts at defining love—that have been made so far. Montagu (1953: 3) perceived love as the most important 'ingredient' in social structures to make us fully human and remarked that we are so used to having a clear definition of a phenomenon at the beginning of an academic inquiry—but it seems that with love it is often the other way round: at the end of a long inquiry process we might come up with a meaningful definition. As Montagu points out, dictionary definitions normally revolve around love being a feeling of deep regard, fondness and devotion. For the sake of this essay it is important to keep in mind that the majority of these definitions are almost exclusively geared at experiences of love between human beings, i.e. most research on love is completely anthropocentric. Obviously the biggest challenge to introducing the topic of love in relation to nature is the widespread notion of love as something romantic happening between two people or exclusively reserved for close family members. Another version of this 'exclusionism' is the idea of spiritual love which has been more or less hijacked by many religious groups.

Fredrickson (2013), widely considered one of the leading researchers in the field of positive psychology, challenges the currently dominating 'love myth' and asks us—just for the sake of a thought experiment—to drop our narrow concepts of love as a special bond, an exclusive, lasting and unconditional commitment or sexual desire. Instead, Fredrickson proposes that love is an interpersonally situated and socially shared experience of one or more positive emotions marked by investment in the well-being of the other, biobehavioral synchronization and mutually responsive action tendencies—which over time may build embodied rapport. Fredrickson talks about "micro-moments of love" that can happen all the time—when we are enjoying a stimulating conversation with friends or colleagues, when the friendly bus driver waits that extra second so that we could hop on the bus or when we are snuggling up to a partner. In our everyday language we actually often use phrases that convey this perception and feeling of synchronization, i.e. we were on the same wavelength or we really clicked.

Fredrickson's research certainly has the potential of debunking many of the exclusionist romantic love myths but it still remains within an anthropocentric framework that only accepts a definition of love as an experience between humans and basically excludes many experiences of love people experience when in/with nature.

4.4 An Ecological Approach to Loving

Anthropologist Kay Milton proposes an expanded view and a more radical perception of emotions in the sense that she lays the groundwork for an ecological approach to emotions—as opposed to the often narrow concept of emotions operating primarily through social relations. In her book Loving Nature Milton examines how science and religion, different concepts of self-hood, our experiences in/with nature, our enjoyment of and identification with nature all shape our emotional attachment with the natural world. Milton points out that the deeply engrained dichotomy emotion/rationality in many western discourses has led to a stigmatization of people who are open about their passionate caring for nature. How intertwined the nature/culture divide is with this dichotomy emotion/rationality becomes obvious when, as Milton (2002: 4) states, one looks at public discourses on nature protection and realizes the message is that "commitments to some things, like trees, landscapes and non-human animals, are emotional, while commitments to other things, like profit and progress, are rational." Given the scope of this essay, I can only refer to a few points that Milton raises.

An ecological approach to love entails questioning the mainstream perception of perception, so to speak. Bateson (1987, 2000) emphasized that what many of us have conceptualized as a human seeing a tree is in systemic understanding actually a whole system tree-human with many feedback loops that the western mind has decided (or rather been trained?) to leave out. This uni-directional framing (a human perceiving a tree) stands in contrast to the "radically participatory ways of knowing" (Wegmann 2012) of, for example, systemic thinkers like Bateson, Buddhists or Deep Ecologists. Nicholsen (2002) emphasizes that this kind of embodied perceptual reciprocity is a key concept in the work of phenomenologist Merleau-Ponty (1964: 167) who was very interested in the experiences of artists. He quotes the painter Paul Klee saying that, "in a forest I have felt many times over that it was not I who looked at the forest. Some days I felt that the trees were looking at me." The Deep Ecology movement has contributed immensely to highlighting the importance of emotional processes when it comes to human-nature-relationships (see e.g. Naess 1989; Macy 2007). Identification with nature is a central experience in Deep Ecology understanding. Milton (2002: 74) explains how this approach departs from the often moralistic environmentalist mainstream perspective: "we act protectively towards nature, not because, for various reasons, we think we ought to, but because we feel inclined to." The Norwegian philosopher Naess (1989) who coined the term 'Deep Ecology' was primarily interested under which conditions humans widen their often narrow perception of 'self'. Fox (1995) talks about 'transpersonal ecology' and argues that the widest and deepest possible identification with other beings makes it possible to transcend the narrow conception of a bounded, autonomous self.

Kay Milton shares a wonderful anecdote (2002: 73) that is a good example of how this sort of enhanced identification and corresponding emotions can even happen during an academic conference. She tells of a conference entitled 'For the love of nature?' where the researchers actively fused their presentations of 'conventional' scientific research with poetry, music, visual arts etc. A talk given by Jane Goodall proved to be the emotional climax of the conference and Milton wondered beforehand how she would "remain composed in this very public arena" as she had never been able to read about Jane Goodall's work with chimpanzees "without dissolving into tears." To her surprise and relief she realized that she had not needed to worry as "the men on each side of me were quietly sobbing. And a more stoic anthropologist colleague commented afterwards that at least half the audience were openly weeping." On another level, this anecdote resonates with Tonkin's (2005: 60) observation that the widespread mantra "the personal should be distinguished from the professional" runs much deeper than being simply a professional rule; she argues that even in places where there is an "ideology of openness" this distinction is often maintained, "suggesting deeper cultural controls on emotional expressivity."

4.5 From Securely Armoured to Seriously Enamoured?

One of Kay Milton's questions (2002, 2005) might be the central (and the most neglected?) question in mainstream environmental academic discourse: Is it a mere coincidence who is actively engaged and concerned with the well-being of nature and who might be more or less indifferent to the current ecological degradation? Interestingly enough, over the past few years there is an increasing amount of research on the importance of worldviews and their connection to sustainable lifestyles and the recognition that emotions seem to play a central role in forming our most fundamental beliefs 'how the world works' which has often been neglected within the environmental discourse. Hedlund-de Witt (2012) provides a comprehensive meta-analysis of the current main approaches/indicators (e.g. New Environmental Paradigm, Connectivity with Nature Scale) that aim at investigating the relationship between worldviews and sustainable behavior. The author highlights the strengths and weaknesses of these survey-based approaches stemming from different disciplines and proposes an Integrative Worldview Framework which focuses on the larger whole instead of looking exclusively at single constructs which often rest upon binaries. The five overarching aspects of the Integrative Worldview Framework are ontology, epistemology, axiology, anthropology and societal vision. Hedlund-de Witt acknowledges that spiritual phenomena like love, awe, wonder and deep reverence for nature have received little attention from researchers, and that this stands in stark contrast to more qualitative research findings and 'common sense' where they are being portrayed as very important.

In his book *Biophilia*, Wilson (1984: 1) describes his biophilia hypothesis as a human "innate tendency to focus on life and lifelike processes." The biophilia hypothesis is enjoying great popularity among environmentalists. However, only a few people are aware that the term biophilia was originally coined and used by the social psychologist Erich Fromm. This disconnect might be explained by what the anthropologist Tonkin (2005: 60) describes as a long history of separation and strict boundary maintenance between anthropology and social psychology and that both have made astonishingly little effort to cooperate in the investigation of the importance of emotions.

Fromm (1973: 406) defined biophilia as "the passionate love of life and of all that is alive; it is the wish to further the growth, whether in a person, a plant, an idea, or a social group. The biophilous person prefers to construct rather than to retain. He wants to be more rather than to have more." For Fromm it was definitely not a coincidence who turned out to be biophilous. He understood biophilia as a state that was "biologically normal", whereas the opposite—necrophilia—was regarded as a psychopathological phenomenon that only occurred if the development of biophilia was stunted.

One of the few researchers who has picked up this seemingly forgotten Fromm-thread is Orr (1993: 416) who poses several important questions: "is biophobia a problem like misanthropy or sociopathy? Or is it merely a personal preference, one plausible view of nature among many? (…) Does it matter that a growing number of people don't like it or like it only in the abstract as nothing more than resources to be managed or as television nature specials?" Orr is correct in pointing out that both, Fromm and Wilson alike, not only view biophilia as innate but basically as a sign of biopsychosocial health.

Orr (1993: 426) muses that "biophilia is a kind of love, but what kind?" and contends that we need to shift from what the Greeks called eros (love of beauty, romantic love aiming to possess) towards agape, which Orr—drawing on Greek philosophy—describes as "sacrificial love that asks nothing in return." Like many other researchers, Orr emphasizes early life experiences and close relationships with caretakers as a central pathway to foster a sense of biophilia. This is confirmed by an enormous amount of data the French physician and researcher Odent (2001) has gathered on how societies that ritually disturb the so-called primal period (from conception until 1st birthday) have had an evolutionary advantage so far because of their potential for aggression and the domination of nature. He draws on anthropological evidence to show cultural settings that have enabled relatively undisturbed physiological processes (e.g. giving birth) but are now more or less extinct. Odent's hypothesis is that the widespread disturbance of perinatal processes (e.g. widely accepted medicalization in many countries) and the resulting disruption of neurobiological priming cascades (e.g. for oxytocin, which has been dubbed the love molecule) lays the early basis for an impaired capacity to love oneself, others and nature. Odent (2001) raises his questions in terms of civilization and even goes so far to ask if humankind can survive modern obstetrics, given the current ecological destruction of our planet.

The anthropologist Turnbull (1978) presents data from a field trip to the Mbuti hunter-gatherers of the tropical forest in north-eastern Zaire. His remarkable introduction bears a resemblance to the writings of many current holistic thinkers like Capra or could even be linked to some of Gregory Bateson's ideas about our inability to perceive the interconnectedness of life.

> What follows may appear to be a needlessly lengthy and overly detailed descriptive account of pregnancy, childbirth and child training. Far from being irrelevant to any study of aggressivity, however, the pity is that I cannot provide more detail. The crux of the learned aspect of aggressivity, at least, is the relationship the individual develops with the world around him. It is a total relationship. We divide it into artificial segments and talk of man's relationship with the human or animal worlds, with the natural and the mechanical or technological worlds, and in many other ways. But for true non-aggressivity and nonviolence to be learned the individual has to gain confidence in his relationship with all the various segments of his experience, and perceive it as a single totality rather than a mere sum total of separate relationships (Turnbull 1978: 161).

Therefore, in my view, the role and importance of embodiment when it comes to our capacity to love (including loving nature) might be worth highlighting. It would exceed the framework of this paper to delve deeper into neurobiological research or interesting studies from the field of epigenetics how our anchor culture literally gets under our skin. However, I would like to highlight some ideas that are similar to

Fromm's ideas on biophilia/necrophilia and how certain life experiences become embodied.

Reich (1950) developed the concept of body armour and character armour. Reich defined body armour as the sum of all chronic, automatic and involuntary muscle tensions that a human being develops in order to defend oneself against their emotions such as anxiety, anger or sexual arousal. Character armour means all psychological attitudes that are used—automatically, subconsciously—to fight off disturbing emotions like anxiety, anger etc. This leads to emotional rigidity, a feeling of inner emptiness and a lack of authentic contact with oneself, others and nature. The process of suppressing and controlling emotions takes place within a socio-political framework that basically favours heavily armoured persons as they are able to function more smoothly in for example authoritarian settings, and will seldom be the ones having problems conforming. The loosening, removing of this armour could not be accomplished by talking alone. Reich came to understand that in many people all this talking about, an excessive intellectualization, was actually a defence mechanism itself. This idea sets him apart from researchers like, for example, Fromm who hoped that psychotherapeutic intervention on a large scale might bring about social change. Reich could see from his extensive practice as a doctor and therapist that even though many people were cognitively able to grasp their defence mechanisms, they felt trapped, they still felt they did not have the capacity to change—because the approach was not holistic, did not include the body. He observed that relatively unarmoured persons shared certain characteristics: they were able to engage in open, loving contact with others, they had a knack for establishing a connection with children, they felt deeply connected with nature and their thinking was less dominated by moralistic binaries of good/evil. These also happen to be some of the main characteristics of biophilious persons, as defined by Erich Fromm.

Going back to the earlier statement by Kay Milton about Deep Ecology and the spontaneous inclination of some people to identify with nature, in Reich's understanding this could be explained by life experiences and how those had become embodied. There is an emerging interest in and recognition that societal change is unlikely to take place without considering our embodied emotions. The peace researcher Wolfgang Dietrich (2012: 251) proposes an idea of 'transrational peaces' and states that peacework (in the broadest sense) cannot ignore the memories of the body—and that a major route to alter psychological structures is through "breath and working on the body's muscle armour."

Reich believed that even if an armoured person wanted to express impulses of love, these were distorted and could not be fully expressed. Above all, the idea of "contactlessness" and the corresponding apathy with regards to fellow human beings and nature is something that seems to be a major obstacle to societal transformation. In a similar way, Wegmann (2012: 83) talks about the "fortress-like self-mode" that many aspects of capitalist/urban life seem to require of us and that block our ways of being with nature.

Applying Milton's (2002: 155) ecological view of emotions and bodily perceptions could broaden this discussion beyond the social-psychological dimension.

For instance, Milton recalls how during her fieldwork in Africa she learned to fear snakes whenever she was walking through tall grass and noticed how her leg muscles tightened strongly. This effect lasted for many months after her return to Ireland even though she knew that there are no snakes to fear. I think these ideas on armouring could actually provide some clues why it is so difficult to change, and to love, even though we might be intellectually aware of certain dysfunctional patterns in our lives.

4.6 Spaces of Resistance and Spaces of Complicity

A few years ago I was working as a physician in a clinic for child- and adolescent psychiatry in Germany. Usually the day kicked off with a so-called morning circle where every kid would share how they were feeling, what had happened during the previous day, and also had the opportunity to ask questions. Although the children were well aware that these gatherings were about them and they were not supposed to ask personal questions about their attending physician, they brought an (understandable) curiosity about the person with whom they share so much personal stuff. One morning curiosity got the better of the group and they demanded to know why I would come to work by bicycle and not by car like all the other doctors. A few months earlier, when the weather was still warm and bike-friendly, they had simply thought I came by bike because I liked sports. Or that I did it for health reasons. I openly admitted that organized sports seemed like a drag to me, that being fit and healthy was a pleasant side effect of cycling and walking, but that my main reason was that I enjoyed being in/with nature and I felt this daily act of resistance was important. It came from a place of loving nature. I shall never forget the silence in the room. One boy wanted to know if I used my car to go to places further away, like Berlin or Hamburg. I laughed. 'No', I answered, "I use the train. And besides, I have never owned a car in my life. I don't even have a driving licence." Now that answer caused some turmoil! I remember the shocked look on the face of the nurse who was present. "But why?" the group pressed on. "For the same reason. I joined Greenpeace when I was a teenager, and let's just say, when love of nature meets certain information and knowledge, it becomes more difficult to participate in certain destructive things."

Why am I sharing this incident? In my view it is a demonstration of how the experience of loving nature and especially small acts of resistance that arise from it seem almost outlandish in the mainstream discourse of affluent global north societies. Had I said I ride my bike because it keeps me healthy and slim, I am sure no one would have found that even remotely odd. But to name as my first and foremost motivation a sense of love and connectedness with nature went against the master narrative, so to speak. Meadows et al. (2004: 281) summarizes this constellation aptly when she writes:

> One is not allowed in the industrial culture to speak about love, except in the most romantic and trivial sense of the word. Anyone who calls upon the capacity of people to practice brotherly and sisterly love, love of humanity as a whole, love of nature and of our nurturing planet, is more likely to be ridiculed than to be taken seriously.

And Milton (2002: 141) invites us to look more closely why for example indigenous activists are ‘permitted’ to talk about their sacred links with the land, while “non-native people might also hold the mountain sacred, but that they feel unable to say so, at least on their own behalf, because they know that arguments about sacredness would not be accepted coming from them.”

Perhaps it is a glimmer of hope that love of nature (as a concept) is allowed more and more into the academic mainstream, maybe even part of the magical “shift of perception” that many Capra (1982, 2002)-inspired academics and activists mention and advocate? But too often we hastily assume that now we ‘have’ a different perception, we finally see the interconnectedness of all things, we see the vulnerability of the web of life and how much our own actions and non-actions have an impact—but into which daily actions do we translate our perceptions and feelings? I may claim that my love of nature also manifests itself in the fact that I decide to not use a car and that I feel happy about it, that I do not perceive it as a sacrifice. Sachs (2010) would call it intelligent self-limitation (a pursuit he highly recommends to the transnational consumer class in general). The Deep Ecologists would say that in fact, I am protecting myself as a tiny part of nature; transhuman love would make no distinction. I am nature and nature is me.

But there is a catch to this wonderful reasoning. I might engage in certain practices of resistance but just as much will I be complicit with the neoliberal nature-destroying machinery that I view so critically. It is like an encounter with a shadow speechlessness. This time the speechlessness is not due to what the mythologist Joseph Campbell called aesthetic arrest, deep awe and wonder when being in/with nature. This time it is a painful speechlessness.

Although I have made certain ecologically minded commitments with regard to my lifestyle, there is no denying that I, as a citizen of the global north, as a part of this transnational consumer class, am participating in the destruction of nature. The emotional drain that comes with this realization, that I am part of the destruction of what I claim to love dearly, is hard to describe. Nicholsen (2002: 7) captures the spirit of this speechlessness well:

> In the urgency of our situation, this speechlessness is mysterious. In hiding the depth of our concern from others, perhaps we also hide it from ourselves. Would it make a difference if we were able to be more courageous in speaking it?

I feel there is a lot of unaddressed pain resulting from my own (and maybe others’) hypocrisy and the manoeuvring to keep the inevitable cognitive dissonance in check. Moreover, it means a heightened awareness of the structural and cultural violence that I am encountering on a daily basis. An example: I turn down an offer to speak at a conference in Minneapolis because I would pollute on such a scale by using an airplane for a 20 min lecture, but the systemic constellation within academia is such that I am being told if I want to ‘get ahead’ there is no space for such reasoning and I should re-think my choices. However, I accept an offer to co-teach a 2 week-course in Costa Rica—which I justify with lots of strategies well-known to reduce cognitive dissonance. The air travel pollution is just as worse as in the first scenario.

Nora Bateson (2012), one of the daughters of Gregory Bateson, said in a recent interview that her father's concept of the double bind has been mainly understood as a theory that is important for psychology. In her view, however, "the double bind we are in is one in which to feed our children and survive from day to day, we are taking part in socio-economic systems that are eliminating the long-term survival of ourselves and our children. We are stuck either way. We are in a feedback loop." This is probably a statement that many people in the global north with immense ecological awareness and love for nature could underwrite. Again, the question arises how to transform this destructive double bind modus? Bateson (2012) goes on to explain that the first step out of this double bind modus is becoming aware of it and making it transparent—to ourselves, to others. And then start searching for creative ways to transcend this destructive pattern. In a sense, my writing about my own hypocrisy and inner tensions and dissonances is a first step towards this sort of meta-communication. In the long run, however, it will not be sufficient if this awareness and re-connection to own feelings does not translate into daily practices of sustainable living.

The current scenario is rather bizarre in the sense that many peace and conflict scholars and peacebuilders engage in a lifestyle that is by no means within the limits of earth' carrying capacity—but go happily about telling other people how to live up to the socioeconomic standards of the global north by adhering to a sustainable development trajectory. Gandhi's famous dictum that the means and ends of any endeavor creating peace should be aligned is often far from the reality many so-called professional peacebuilders.

Biersack (2006: 28), drawing on the work of Michael Herzfeld, argues for a "militant middle ground" in the arena of political ecology which remains "grounded in an open-ended appreciation of the empirical." I think this could prove a useful metaphor for not getting caught up in these two spaces that I have outlined here, and might even be a place from where to gather new strength. It also resonates with the idea of "planetary love", put forward by Mickey/Carfore (2012: 122), who have been inspired by the work of ecofeminist scholars like Haraway, Spivak and Plumwood. Planetary love is a force that "rather than affirming or opposing the globalization, acknowledges its complicity with the neocolonial tendencies of globalization while aiming toward another globalization." Another globalization is possible—but who is going to set the coordinates when it comes to limits?

4.7 Who Defines and Measures What the Limit Is?

The word unlimited is a big one. Unlimited love? Unlimited nature? Looking at the current global situation both scenarios seem difficult to imagine. Escobar (1999: 14) asks if "nature can be theorized within an anti-essentialist framework without marginalizing the biological?" Given the contemporary popularity of poststructuralist theories in the academic world and Escobar's own (anti-essentialist-based) useful differentiation between organic nature, capitalist nature and techno-nature, this appears to be a major question. In fact, authors like Latour (2004: unpaginated) have highlighted the potential devastation that arises when this question does not get more attention: "dangerous extremists are using the very same argument of social construction to destroy hard-won evidence that could save our lives (…) Why does it burn my tongue to say that global warming is a fact whether you like it or not?"

The idea of a biophysical limit/reality would imply that there is some sort of essence regarding nature, albeit this essence itself might be more fluid and changeable than we can currently grasp. But the question remains: if there is a biophysical limit, who defines what the limit is? How are we going to frame certain measurements, quantitative data? And how is this connected to the concept of loving nature?

As happens frequently, almost every phenomenon that is officially admitted into academic mainstream is at one point submitted to the exercise of being quantified/measured. Loving nature is no exception. Attempts to measure love and care for nature (Perkins 2010) and significant positive correlations between love of nature and well-being, mental health or mindfulness (see e.g. Gullone 2000; Howell et al. 2011; Cervinka et al. 2012) have been reported. Although these constructed measurement scales can be easily critiqued as anthropocentric shallow ecology, they bear witness to an increasing shift towards ecocentric notions of love, and could actually function as some sort of 'door opener' to expanding the current scientific research on love.

Why is this important? In my view, many global north conceptions of love (especially romantic love), consumption and consumerist culture are intimately linked. In the spirit of Transition Town Movement founder Hopkins (2008), I would argue that a majority of people in the global north has forgotten how to have a good time with little consumption. Few people would object to flying thousands of air miles for a 'romantic weekend'—because it is the ultimate proof that your partner loves you, right?! And even 'loving nature' is packaged into a scenario of mass tourism or maybe even chic eco-tourism and often functions as a mere background for romantic human love to unfold.

But how could we define 'sustainable love' then? Many transnational consumer class members are blissfully unaware that despite their wonderful gardens, visits to nature parks, being enchanted with nature TV programs etc. their lifestyle is exceeding earth' carrying capacity by far. I think we might get some inspiration from looking at the current interest in happiness and well-being in many disciplines. Despite my rather poetic outlook on life, I am aware that the current interpretational hegemony is constantly supported and nourished by data and indicators that convey a message of how well things are going in the global north. I believe that in order to create counter-narratives that actually have an impact on the current master narrative, the creative capture of quantitative data might be indispensable.

In 2006 the London-based New Economics Foundation published their so-called Happy Planet Index for the first time. Interestingly enough, the two parameters that were more or less absent from orthodox indicators of progress at this time—human well-being and ecological sustainability—happen to be the central parameters of the Happy Planet Index.

The Happy Planet Index (Abdallah et al. 2012) combines the calculated Happy Life Years (which are based on subjective life satisfaction and the average life expectancy at birth) with the Ecological Footprint (Wackernagl/Rees 1996; Wackernagl 2013) of the respective nation. Put simply: the Happy Planet Index gives us a fairly good idea which countries' inhabitants manage to lead happy lives without exceeding earth' ecological carrying capacity. Or as the New Economics Foundation puts it: good lives don't have to cost the earth! The current research on happiness and well-being persistently points to the importance of loving relationships and I think that a Happy Planet would certainly put loving relationships in a central spot.

To fully understand the significance of the work that the New Economics Foundation has undertaken, it is worth taking a moment to reflect which countries we hear mentioned most frequently in the mainstream media as being 'successful', 'powerful', 'peaceful' and 'highly developed'. Among the countries mentioned and praised to the skies—and backed up by numerous mainstream indicators—we will find for example countries like Switzerland, Norway, Denmark, Iceland, the US, Japan, Germany. While many of these countries score among the highest for life satisfaction, GDP, life expectancy, Happy Life Years, Human Development Index etc. they share one serious flaw—their national Ecological Footprint exceeds earth's carrying capacity by a long way, so their Happy Planet Index score is on the rather unhappy side.

My current host country, Canada, is a good example. In terms of indicators for well-being and life satisfaction, Canada lives up to its reputation as an immigrant magnet due to high levels of comfort and well-being: out of 151 countries Canada ranks number 2 when it comes to subjective well-being. In terms of life expectancy, Canadians, with an overall life expectancy of 81.0 years, land on place number 12 out of 151 ranked countries. Keep in mind that this is the moment when orthodox economists like to remind you that therefore the Canadian model is working just fine and is desirable for other countries to imitate. Now we figure in the Ecological Footprint (the third component of the Happy Planet Index) and this changes the whole picture drastically since Canada has currently the 8th largest Ecological Footprint per capita in the world—if everyone lived as the average Canadian we would need 3.5 planets to sustain life. Consequently, Canada ranks only number 56 out of 151 countries in the Happy Planet Index (Abdallah et al. 2012). This scenario is more or less the same for all the seemingly well-to-do, happy nations such as Switzerland, Denmark or Iceland. Their supposed happiness and peacefulness is simply not ecologically sustainable.

The Happy Planet Index certainly does not measure everything that is important and has its own shortcomings—but my point is that countries which manage to achieve a high life satisfaction without destroying our earth, without insatiably accumulating tons of unnecessary and unsustainable goods deserve much more of our (research) curiosity. The New Economics Foundation has created an immense awareness for this concept during the past years by providing this index. And I think that it is a valuable tool for the deconstruction of the often rather naïve and gullible conception of loving nature in the global north.

And finally, Mathews (2002: 222) reminds us that the trap for environmentalists has been to conceive of nature in terms of things instead of processes. In her understanding a return to nature does not entail to "restore a set of lost things or attributes, but rather to allow a certain process to begin anew. This is the process that takes over when we step back, when we cease intervening and making things over in accordance with our own abstract designs." Obviously, this "ethos of countermodernity" does not seem to go well with the modern/postmodern busyness of many peacebuilders. But I think it resonates with an idea of loving put forward by Maslow (1966: 116) who said that ": if you love something or someone enough at the level of Being, then you can enjoy its actualization of itself, which means that you will not want to interfere with it, since you love it as it is in itself."

4.8 Conclusion

During the past decade the 'scientification of love' (Odent 2001) has led to an increased openness and willingness to allow love as a research topic to enter the scientific databases. Since many people in western countries tend to associate love with a rather narrow, romantic scenario an ecological approach to love, as proposed by anthropologists like Milton (2002), could provide an important entry point to make visible a few dangerous blind spots concerning the current anthropocentric models of love.

Moreover, the discourse on the biophysical reality/limits of nature might benefit from engaging with newly emerging indicators like the Happy Planet Index (Abdallah et al. 2012) as to include the happiness/well-being dimension vis-à-vis quantitative data on sustainable consumption patterns. Put bluntly, loving (or at least respecting) nature and acting accordingly appears to be a prerequisite for love between humans at this point in time. The current global ecological degradation reminds us that focusing on human-human aspects of love alone tends to neglect the simple fact that we are destroying what gives us life—while being proud of our loving behavior towards other human beings.

Given some of the presented research findings, it seems that developmental pathways in early life and how experiences become stored in a bodily sense, might play a greater role in determining the human capacity to love nature (or to love in a general sense) than the field of peace and conflict studies does currently acknowledge. The challenge for the emerging ecological approach to love might be to take a closer look at the deep-psychological and neurobiological dimensions of apathy, alienation and destructive aggression. There is a certain preference for the less disturbing *biophilia* model made popular by Wilson, but this may be due to the lack of interdisciplinary borrowing from critical socio-psychological theories and the emerging research on embodiment.

Ultimately, there is a certain speechlessness when it comes to experiences of love, of deep wonder when being with/in nature. This makes it difficult to write about these important phenomena. In fact, Bateson (1987) suggested that experiences of sacredness seem to rely on some information remaining hidden and not explicitly communicated. Equally difficult for many global north citizens: to address our own hypocrisy and complicity concerning the destruction of nature and find ways to re-connect to these fragmented emotional particles and change our daily life practices. But as one of the participants of the Deep Ecology conference, which I mentioned at the beginning of this essay, said: "in the end, we human beings are quite simple. We care for and protect what we love."

References

Abdallah, Saamah; Michaelson, Juliet; Shah, Sagar; Stoll, Laura; Marks, Nic, 2012: "The Happy Planet Index", in: *New Economics Foundation, Report 2012*. Last modified June 2012; at: http://www.happyplanetindex.org/assets/happy-planet-index-report.pdf.

Adorno, Theodor W., 1997: *Aesthetic Theory* (Minneapolis: University of Minnesota Press).

Anielski, Mark, 2007: *The Economics of Happiness: Building Genuine Wealth* (Gabriola Island, BC: New Society Publishers).

Bateson, Gregory, 2000: *Steps to an Ecology of Mind: Collected Essays in Anthropology, Psychiatry, Evolution and Epistemology* (Chicago: University of Chicago Press).

Bateson, Gregory; Bateson, Mary Catherine, 1987: *Angels Fear: Towards an Epistemology of the Sacred* (New York: Macmillan).

Berglund, Eeva, 2006: "Ecopolitics Through Ethnography: The Cultures of Finland's Forest-Nature", in: Biersack, Aletta; Greenberg, James B. (Eds.): *Reimagining Political Ecology* (Durham and London: Duke University Press): 97–120.

Biersack, Aletta; Greenberg, James B. (Eds.), 2006: *Reimagining Political Ecology* (Durham–London: Duke University Press).

Capra, Fritjof, 1982: *The Turning Point. Science, Society and the Rising Culture* (Toronto: Bantam Books).

Capra, Fritjof, 2002: *The Hidden Connections: Integrating the Biological, Cognitive and Social Dimensions of Life into a Science of Sustainability* (New York: Anchor Books).

Cervinka, Renate; Röderer, Kathrin; Hefler, Elisabeth, 2012: "Are Nature Lovers Happy? On Various Indicators of Well-Being and Connectedness With Nature", in: *Journal of Health Psychology*, 17,3: 379–388.

Charlton, Noel G., 2008: *Understanding Gregory Bateson: Mind, Beauty and the Sacred Earth* (Albany, NY: State University of New York Press).

Denskus, Tobias, 2007: "Peacebuilding Does not Build Peace", in: *Development in Practice*, 17,4/5: 656–662.

Drengson, Alan; Inoue, Yuichi, (Eds.), 1995: *The Deep Ecology Movement: An Introductory Anthology* (Berkeley, CA: North Atlantic Books).

Escobar, Arturo, 1999: "After Nature: Steps to an Antiessentialist Political Ecology", in: *Current Anthropology*, 40,1: 1–30.

Escobar, Arturo, 2011: "Sustainability: Design for the Pluriverse", in: *Development*, 54,2: 137–140.

Fredrickson, Barbara, 2013: *Love 2.0.: How Our Supreme Emotion Affects Everything We Feel, Think, Do and Become* (New York, NY: Hudson Street Press).

Fromm, Erich, 1973: *The Anatomy of Human Destructiveness* (New York: Picador).

Fox, Warwick, 1995: *Toward a Transpersonal Ecology: Developing New Foundations for Environmentalism* (New York: State University of New York Press).
Gullone, Eleonora, 2000: "The Biophilia Hypothesis and Life in the 21st Century: Increasing Mental Health or Increasing Pathology?", in: *Journal of Happiness Studies*, 1: 293–321.
Hedlund-de Witt, Annick, 2012: "Exploring Worldviews and Their Relationships to Sustainable Lifestyles: Towards a New Conceptual and Methodological Approach", in: *Ecological Economics*, 84: 74–83.
Hinds, Joe; Sparks, Paul, 2008: "Engaging with the Natural Environment: The Role of Affective Connection and Identity", in: *Journal of Environmental Psychology*, 28: 109–120.
Hopkins, Rob, 2008: *The Transition Handbook: From Oil Dependency to Local Resilience* (White River Junction, VT: Chelsea Green Publishing Company).
Hopkins, Rob, 2011: *The Transition Companion: Making Your Community More Resilient in Uncertain Times* (White River Junction, VT: Chelsea Green Publishing Company).
Howell, Andrew J.; Dopko, Raelyne L.; Passmore, Holli-Anne; Buro, Karen, 2011: "Nature Connectedness: Associations with Well-Being and Mindfulness", in: *Personality and Individual Differences*, 51: 166–171.
Ingold, Tim, 2011: *Being Alive: Essays on Movement, Knowledge and Description* (London–New York: Routledge).
Jónasdóttir, Anna G.; Ferguson, Ann (Eds.), 2014: *Love: A Question for Feminism in the Twenty-First Century* (New York–London: Routledge).
Kellert, Stephen R.; Wilson, Edward O. (Eds.), 1993: *The Biophilia Hypothesis* (Washington DC: Island Press).
Latour, Bruno, 2004: "Why Has Critique Run Out of Steam? From Matters of Fact to Matters of Concern", in: *Critical Inquiry*, 30,2: online edition.
Macy, Joanna; Brown, Molly Young, 1998: *Coming Back to Life: Practices to Reconnect Our Lives, Our World* (Gabriola Island, BC: New Society Publishers).
Macy, Joanna, 2007: *World as Lover, World as Self: Courage for Global Justice and Ecological Renewal* (Berkeley, CA: Parallax Press).
Maslow, Abraham H., 1966: *The Psychology of Science: A Reconnaissance* (New York–London: Harper & Row Publishers).
Mathews, Freya, 2002: "Letting the World Grow Old: An Ethos of Countermodernity", in: Schmidtz, David; Willott, Elizabeth (Eds.): *Environmental Ethics: What Really Matters, What Really Works* (Oxford & New York: Oxford University Press).
Maturana, Humberto Romesin; Verden-Zöller, Gerda, 2008: *The Origin of Humanness in the Biology of Love* (Exeter: Imprint Academic).
Meadows, Donella; Randers, Jorgen; Meadows, Dennis, 2004: *Limits to Growth—The 30 Year Update* (Chelsea: Green Publishing Company).
Merleau-Ponty, Maurice, 1964: *The Primacy of Perception* (Evanston, IL: Northwestern University Press).
Mickey, Sam; Carfore, Kimberly, 2012: "Planetary Love: Ecofeminist Perspectives on Globalization", in: *World Futures: The Journal of Global Education*, 68,2: 122–131.
Milton, Kay, 2002: *Loving Nature: Towards an Ecology of Emotion* (London–New York: Routledge).
Montagu, Ashley, (Ed.), 1953: *The Meaning of Love* (New York: The Julian Press Inc.).
Naess, Arne, 1989: *Ecology, Community and Lifestyle: Outline of an Ecosophy* (Cambridge: Cambridge University Press).
Nicholsen, Shierry Weber, 2002: *The Love of Nature and the End of the World: The Unspoken Dimensions of Environmental Concern* (Cambridge, MA: The MIT Press).
Odent, Michel, 2001: *The Scientification of Love*, 2nd edn (London: Free Association Books).
Orr, David W., 1993: "Love It or Lose It: The Coming Biophilia Revolution", in: Kellert, Stephen R.; Wilson, Edward O. (Eds.): *The Biophilia Hypothesis* (Washington DC: Island Press): 415–440.

Pálsson, Gisli, 2006: "Nature and Society in the Age of Postmodernity", in: Biersack, Aletta; Greenberg, James B. (Eds.): *Reimagining Political Ecology* (Durham–London: Duke University Press): 70–93.

Pease, Bob, 2012: "Interrogating Privileged Subjectivities: Reflections on Writing Personal Accounts of Privilege", in: Livholts, Mona (Ed.): *Emergent Writing Methodologies in Feminist Studies* (New York & London: Routledge).

Perkins, Helen E., 2010: "Measuring Love and Care for Nature", in: *Journal of Environmental Psychology*, 30: 455–463.

Pyle, Robert Michel, 2003: "Nature Matrix: Reconnecting People and Nature", in: *Oryx*, 37,2: 206–214.

Reich, Wilhelm, 1950: *Character Analysis*, 4th edn (London: Nevill Vision).

Sachs, Wolfgang, (Ed.), 2010: *The Development Dictionary: A Guide to Knowledge as Power*, 2nd edn (London: Zed Books Ltd).

Tonkin, Elizabeth, 2005: "Being There: Emotion and Imagination in Anthropologists' Encounters", in: Milton, Kay; Svasek, Maruska (Eds.): *Mixed Emotions: Anthropological Studies of Feelings* (Oxford: Berg Publishers): 55–69.

Turnbull, Colin M., 1978: "The Politics of Non-aggression", in: Montagu, Ashely (Ed.): *Learning Non-aggression. The Experiences of Non-literate Societies* (Oxford: University Press): 161–221.

Wackernagel, Mathis; Rees, William. E., 1996: *Our Ecological Footprint: Reducing Human Impact on the Earth* (Gabriola Island, BC: New Society Publishers).

Wegmann, Regula, 2012: "Loving Nature" Nature's Way: Exploring Radical Participation with Nature Through the Metaphor of Complex, Dynamic Self-systems", in: *World Futures: The Journal of Global Education*, 68,2: 82–92.

Wilson, Edward O., 1984: *Biophilia: The Human Bond with Other Species* (Cambridge, MA: Harvard University Press).

Other Literature

Abdallah, S.; Michaelson, J.; Shah, S.; Stoll, L.; Marks, N., 2012: "The Happy Planet Index. Report 2012"; at: www.happyplanetindex.org (10th Oct 2013).

Bateson, N., 2012: "An Ecology of Mind: Film Interprets Life of Unconventional Thought", at: http://www.theecologist.org/campaigning/culture_change/1198372/an_ecology_of_mind_film_interprets_a_life_of_unconventional_thought.html.

Wackernagel, Mathis, 2013: "Global Footprint Network", at: www.footprintnetwork.org/en/index.php/GFN/page/footprint_basics_overview/.

Chapter 5
Drowning in Complexity? Preliminary Findings on Gender, Peacebuilding and Climate Change in Honduras

Henri Myrttinen

Abstract Although the interconnected issues of gender, climate change and peacebuilding have been high on the international agenda for the past decades, and the interplay between the issues has been researched in pairs (i.e. gender/peacebuilding, gender/climate change, climate change/peacebuilding), there is little research available on the simultaneous interplay of the three. This chapter examines first some of the theoretical issues and dynamics at play when addressing these issues, followed by an overview of preliminary findings from Honduras. These findings are based on scoping research carried out by and for International Alert, in preparation for a project on the gendered impacts of climate change-induced coffee rust and peacebuilding approaches to help communities cope with these.

Keywords Gender · Peacebuilding · Climate change · Honduras · Coffee rust

5.1 Introduction

At least for the past decade and a half, since the establishment of the United Nations Security Council Resolution 1325 (2000) on Women, Peace and Security mandate, there has been a realisation at the policy level that gender needs to be taken into account when dealing with conflict resolution and peacebuilding. Gender, climate change adaptation and sustainable development have previously been linked in development literature and explicitly stated, for example, in IPCC (2001). The interlinked nature of peace and development were also at the centre of peacebuilding debates throughout the 2000s and onwards (World Bank 2011a). Clearly, then, gender, climate change and peacebuilding are linked to each other. However, while pairing the issues—gender/peacebuilding, gender/climate change, climate change/peacebuilding—has been relatively straightforward, what has proven more

Dr. Henri Myrttinen is the Head of the Gender Team at International Alert. He holds a M.Sc. in environmental engineering from the Tampere University of Technology, Finland, and a Ph.D. in Conflict Resolution and Peace Studies from the University of KwaZulu-Natal, South Africa; Email: hmyrttinen@international-alert.org.

H.G. Brauch et al. (eds.), *Addressing Global Environmental Challenges from a Peace Ecology Perspective*, The Anthropocene: Politik—Economics—Society—Science 4, DOI 10.1007/978-3-319-30990-3_5

difficult has been working on the three issues simultaneously, although some initial steps have been taken (see for example United Nations 2013). What has been especially challenging has been to do so without, on the one hand, generalising to the point where one remains at the level of banal platitudes and, on the other, drowning in the complexity of local specificities.

In this chapter, I will attempt to first map some of the conceptual issues surrounding gender, climate change and peacebuilding, followed by providing a case example of a project currently being developed in Honduras by International Alert and partner organisations that aims to address all of these issues. Based on some initial findings coming out of the research, I will then seek to draw on some preliminary findings.

International Alert is a London-based, non-governmental peacebuilding organisation which has long been working both on integrating gender perspectives and climate change into its and others' work on ending and preventing violent conflict.[1] The findings presented here on Honduras are based both on desktop research and initial research conducted by the author in the country, as well as an unpublished study conducted by two local researchers commissioned by Alert. Both the initial research and the unpublished study are part of the preparatory groundwork for a future project on addressing the gendered impacts of climate change from a peacebuilding perspective. At the time of writing, the project is still in its inception phase but is due to commence in 2015, with the prospective main focus of the work likely to be in the Departments of Copán and Santa Bárbara.

5.2 Why Gender?

Gender—defining ourselves as either women, boys, girls, men, transgender or intersex persons—is often a key factor in determining social spaces and opportunities available or not available to us, thereby setting the parameters of vulnerability. Gender is, however, not the only factor in determining vulnerability: neither women nor men are a homogenous category with the same needs and possibilities, nor do all of the needs and concerns of sexual and gender minorities overlap. Rather, gender needs to be seen in conjunction with other social identity markers, such as social class, age, sexual orientation, marital status, disability, urban/rural location and, at times, ethnic or religious background and caste.

Gender roles and expectations are also situational—and time-bound—both in the historical sense in the way that, say, Victorian-age British middle-class femininities differ from those of today and in a more everyday sense, i.e. an office worker will display a different kind of masculinity nine to five compared to after work in the pub with his mates. While these different factors in many ways determine the broad framework in which we are able to perform our various gender identities as well as

[1]For more information on the organisation, see at: http://www.international-alert.org.

possible vulnerabilities, personal agency should not be overlooked—although the various social markers also often determine in how far one is able to have agency.

Both in violent conflict and in disasters—be they short- or long-term—gender is a key co-determinant of vulnerability and opportunity. In conflict, for example, certain men (mostly young, often lower class) are socially conditioned, both by other men and also by their mothers, girlfriends, wives, to use the physical capital of their bodies either for protection or for their personal and/or communal benefit, e.g. by acting as soldiers, gangsters, guerrillas, cattle raiders or the like (see e.g. Wright 2014). Sometimes they are joined by women in their enterprises, but more often women and other men provide the social, moral and material support without which they could not act violently.[2]

As Ormhaug et al. (2009: 2–3) point out, quantitative data on how violent conflict impacts on women and men is difficult if not impossible to come by: "there are practically no global data available that allowed us to investigate conflict mortality disaggregated by gender", although, based on the data, they were able to gather that "men are more likely to die during conflicts, whereas women die more often of indirect causes after the conflict is over" (see also Urdal 2010). Physical casualties are of course not the only victims of conflict: for those displaced, those caring for wounded family members or those carrying the emotional scars of conflict, gender plays a key role in determining access to resources, vulnerability and also socially accepted ways of dealing with their suffering. More important in my opinion than the quantitative data, however, are the qualitative differences: for example, gender, age, class and other factors increase the risk of certain men facing violent battlefield death compared to other demographic groups. Other groups may, for example, be far more vulnerable to *sexual and gender-based violence* (SGBV).

As for conflict, disaster affects different men and women in a variety of ways. Some challenges will be similar, but gendered access to resources and political decision-making processes, or exposure to violence, including SGBV, also mean different vulnerabilities for different groups. As Neumayer/Plümper (2007: 551) put it:

> […] a vulnerability approach to disasters would suggest that inequalities in exposure and sensitivity to risk as well as inequalities in access to resources, capabilities and opportunities systematically disadvantage certain groups of people, rendering them more vulnerable to the impact of natural disasters.

Here it is then also necessary to look beyond broad, supposedly homogenous gender categories and look at the needs and social positionalities of, say, young,

[2]On the multiplicity of the roles played by women in conflict-affected situations, see for example Cohn (2012) and Myrttinen et al. (2014). Although the majority of active combatants in the world's conflicts are men and boys, female combatants sometimes make up a considerable part of the fighting forces, for example, an estimated 20–40 % of the Colombian *Fuerzas Armadas Revolucionarias de Colombia* (FARC) (Herrera/Porch 2008) and up to 30–40 % of the former Maoist *People's Liberation Army* (PLA) in Nepal (Roy 2008; Yami 2007).

urban, lower-class men compared to older, rural, upper-class women, and what these mean in terms of their needs and possibilities. Looking seriously at gender also requires overcoming heteronormative assumptions, in at least two ways (see also Jauhola 2013). Firstly, outside interveners need to let go of assumptions that all or most households in post-conflict, post-disaster and attendant situations of displacement conform to a nuclear or extended family unit based on a father, a mother and several children. In reality, households are often far more complex and fluid, and this needs to be taken into account in order for policy and programming to effectively meet the needs of the populations affected. Secondly, both in post-conflict and post-disaster programming, the particular needs of registered or, more usually, unregistered same-sex couples and of transsexual, transgender and intersex persons require far more attention than they have received to date (Knight 2014), particularly in situations of displacement.

5.3 Gender-Relational Peacebuilding

The case for approaching peacebuilding from a gender angle is made, for example, in the Programming Framework of International Alert (2010a: 19):

> Gender is one of the factors that influence, positively and negatively, the ability of societies to manage conflict without resorting to violence. Since gender analysis can help us understand complex relationships, power relations and roles in society, it is a powerful tool for analysing conflict and building peace. [...]
>
> There is a relationship between gender relations and continuing cycles of violence. Norms that promote narrow, uncompromising, and violent identities for boys and men are an important underlying cause of high-levels of violence at all levels of society. Individuals who have the courage to break prevailing gender norms and stand up against violence risk losing fundamental rights and endanger their own safety. These dynamics have implications both for conflict analysis and for designing peacebuilding strategies: if gender relations have been a factor in perpetuating violence, they can also be transformed into a strategy for rebuilding more peaceful social relations.
>
> At the same time, violent conflict is a driver for changes in gender relations. In many cases, women have taken on a broader range of economic and societal roles in times of conflict. Conflict can also give rise to more rigid gender stereotypes that men and women are expected to fulfil. Alert should aim to make use of positive changes in gender relations during conflict to promote more peaceful and inclusive societies.

In order to promote gender-responsive peacebuilding with the aim of increasing societal inclusivity, more equitable relationships and reducing vulnerability, be it to conflict or disaster, policy and programming need to be informed by an understanding of the particular local social, economic, cultural and political setting.

A thorough gendered analysis of power dynamics, needs and particular vulnerabilities should form the starting point of interventions, rather than having 'gender' (or, for that matter, "climate change") merely as a box to tick at the end of a donor report.[3] One starting point for examining gender in a comprehensive manner is to view it as a relational dynamic, that is to, say, view masculinities and femininities as being co-produced and defined by men and women in relation to each other and in relation to other social identity markers (see also El-Bushra 2012; Myrttinen et al. 2014).[4]

The challenge is then how to deal with this complexity without drowning in it—if the various age, class, ability, power, ethnic, religious, geographical, temporal, economic, political, social and so on need to be taken into account for various groups of women, men, boys, girls and sexual and gender minorities, how does one manage that?[5] And if it is an analysis of all of the above, where does that leave gender? I will seek to add more concrete substance to the idea of gender-relational peacebuilding by examining what this means in terms of vulnerability to the impacts of climate change more generally, before looking at the specific case of Honduras.

5.4 Vulnerabilities, Climate Change, Gender and Peacebuilding

Equitable access to economic resources are a key component of social inclusion and therefore to peacebuilding, and sustainability is key to ensuring that these resources continue to be available to current and future generations. Climate change, however, is placing increased stress on resources and communities dependent upon them, especially in a range of already fragile and conflict-affected states, thus potentially jeopardizing peacebuilding processes or exacerbating conflicts (see e.g. Peters/Vivekananda 2014; Vivekananda et al. 2014).[6]

Gender, along with other social factors, is often a major determinant in mediating access to resources, land, jobs (some of which are socially 'coded' masculine, some feminine), information and networks of patronage. Many of these tend to be stacked against women more generally and, in particular, against more vulnerable groups such as widows, single women or abandoned wives. They are also stacked against the young and very old, the lower classes, the rural population, the

[3]Sometimes these are conflated, with one box to be filled out by donor funding recipients on how their project "impacted on gender, environment and climate change".

[4]Masculinities are thus defined in part in relation to femininities, heterosexuality is defined in relation to homosexuality, and so on.

[5]All the more so as donors are increasingly pushing for quicker impacts, bigger projects, more "value for money" (i.e. less time and resources for research), 'up-scaling' and replicating of 'what works' (or what was thought to work in a very specific, given context) and of course local ownership, even if the ready-made 'solutions' are being imported from elsewhere.

[6]Given the inevitable complexities of such processes, one should avoid simplistic assumptions that automatic or inevitable linear causality along the lines of climate change leads to resource degradation, resource degradation to increased inequality, inequality to conflict.

displaced, the sexual and gender minorities, often against particular ethnic or religious groups and especially against those with disabilities. Many people thus find themselves struggling with not only one but multiple forms of discrimination.

Vulnerability is, however, not only an issue of large-scale economic and political structural issues, larger socio-cultural norms or the role of external interventions. It can also be exacerbated by seemingly small, mundane manifestations of these disparities: do women and girls have access to telecommunications, are they allowed to open bank accounts without the approval of male relatives, can they evacuate from an area under threat or are they expected to wait for a male relative, are their nutritional needs jeopardized by expectations of men and boys eating meals first, with only leftovers for them?

In many societies, the public and private spheres as well as productive and reproductive sectors of the economy tend to be differently gendered. At the risk of making a vast generalisation, women tend to be expected to be more in the private and men in the public sphere, with women undertaking both reproductive and productive tasks, and men, as presumed primary breadwinners, undertaking paying tasks in the productive sector. These often tend to be the idealized roles. In practice, especially among lower socio-economic classes, and in situations of conflict, disaster, displacement and migration, de facto roles are very different (see also Gutmann 2006). Nonetheless, and in spite of real-life experiences to the contrary, these 'ideal' roles live on as the ones to aspire to (see also Turner 1999).

Dolan (2002: 64) has noted that in the "context of on-going war, heavy militarization and internal displacement it is very difficult, if not impossible for the vast majority of men to fulfil the expectations contained in the model of masculinity" prevalent as the idealized form in society. Nonetheless, for his case study area of northern Uganda, Dolan (2001: 11) argues that

> the normative model of masculinity [...] exercises considerable power over men, precisely because they are unable to behave according to it, but cannot afford not to try to live up to it. The relationship between the social and political acceptance which comes from being seen to conform to the norm, and access to a variety of resources, is a critical one in a conflict situation.

The 'thwarting' of men's gender identities by a yawning gap between aspirations and reality can be further, wittingly or unwittingly, exacerbated by outside interventions. In many ways, external interventions (e.g. humanitarian aid) can 'infantilize' men who are expected to, and expect themselves to, be breadwinners by turning them into agency-less aid recipients. Not being able to support themselves economically can often leave young men trapped in a 'social moratorium' of being perpetually considered 'youth' (and thereby excluded from social and political decision-making processes), in spite of their generational age (Vigh 2006). State agencies, *non-governmental organizations* (NGOs) and international donor agencies also often have a tendency to, in the name of gender equality, focus on women as agents of change and demonise local masculinities as lazy, violent, irresponsible and culturally backward. While the promotion of women's rights and changing of harmful masculine behaviours is extremely important, doing so in a non-conflict, sensitive manner can lead to frustration and to backlash from men, which may put women at a higher risk of violence (International Alert 2010b).

In the societal upheaval of short- or long-term conflict/disaster, displacement and post-conflict/post-disaster, new opportunities are opened up but also new pressures are placed on women and girls to increase their participation in the economic and political spheres, although their vulnerabilities may often also rise (United Nations 2013). Gendered ideologies, however, are also often as slow to change for women as they are for men, and increased spaces, empowerment and new roles are often questioned by both women and men in the aftermath of conflict/disaster as communities seek a return to an imagined patriarchal, heteronormative "golden age" they had prior to the upheaval caused by conflict and disaster (El-Bushra et al. 2014; Jauhola 2013). The aim of gender-responsive peacebuilding is thus to assist communities in finding their own non-violent approaches to dealing with these processes of upheaval, while not exacerbating gendered vulnerabilities but rather promoting more gender equitable power dynamics.

5.5 Honduras—Coffee, Conflict and Climate Change

In this section I will attempt to outline what work on gender, conflict/peacebuilding and sustainability in the face of climate change-induced stress on rural communities might look like, based on preliminary scoping for a future project focusing on these issues in Honduras. The focus country is not a traditional conflict region but one that is nonetheless facing major urban and rural violence, as well as the threat of massive job losses in one of its main economic sectors due to climate change. The possible gendered micro-dynamics of potentially massive, climate-change-related job losses in the coffee sector on rural communities and—through gendered and age-related patterns of internal and transnational migration—on urban and rural communities have not been explored extensively. The project is, at the time of writing, at a preliminary stage and therefore the section will have a more exploratory tone.

5.5.1 Migration

Honduras currently has a population of approximately 8.1 million, and according to World Bank (2011b) data, ranked eighth highest globally on inequality, based on the Gini index, and is the fourth poorest country in the hemisphere. An estimated 700,000 to one million Hondurans live abroad, the majority of which are in the United States of America (USA), with a large number of undocumented migrants. According to Reichman (2013), the migrant population is approximately 53 % male and 47 % female. Deportations from the USA are not uncommon, with 25,000 Hondurans repatriated from the US in 2009 alone (Hirsch 2010). In 2013–2014, there has been a sudden up-turn in the number of unaccompanied minors from

Central America crossing into the USA, with a large number coming from Honduras, often citing widespread gang violence as a motivating factor for risking the perilous journey (UNHCR 2014). The spike of over 50,000 children has led to increased concern about migration in the US and has already led to increased border security not only on the US-Mexican border but also on Mexico's borders with its southern neighbours (WOLA 2014).

Honduran migration patterns are gendered, with migration to the US, in part due to the dangers involved in the crossing, mostly being traditionally seen as a male undertaking (Reichman 2011). Increasingly, however, Honduran women are migrating as well, with Spain being a preferred destination according to anecdotal data.[7] The different migration patterns have the potential to change gendered socio-economic and demographic patterns, as male migration to the US tends to be more circular and shorter-term than European migration and women working in the service sector in Spain are earning more than men doing menial jobs in the US. The jobs available in the European Union (EU) would also seem to be, based on anecdotal evidence, more appealing than those available to Honduran men north of the Rio Grande.[8] A knock-on effect of the availability of only low-skilled jobs for Hondurans in the US has been a reduction in the value placed on education among men (Reichman 2011). Whether or not this is the same for men and women migrating to other countries with the potential for higher-skilled labour will be examined as part of the research.

5.5.2 *Forms of Urban and Rural Violence*

Although Honduras has not been party to an active conflict since the 1969 'Football War' against El Salvador, it can hardly be considered a society at peace. Urban and rural violence have led to the country having the highest per capita homicide rates on the planet and its industrial capital, San Pedro Sula, gained the unenviable epithet "most violent city in the world" (CCSPJP 2012) with a homicide rate of 173 per 100,000 residents.[9]

Much of the urban violence is blamed on gangs (*maras*), especially the transnational *Calle* (or *Barrio*) *18* and *Mara Salvatrucha* gangs. The groups originated in Los Angeles and spread to Central America through the deportation of gang members from the US. Much of the gang violence is targeted against other

[7]Interviews, Tegucigalpa, July–August 2014; the available 2013 data from the Spanish *Instituto Nacional de Estadística* (INE 2014) does not allow for a confirmation of this anecdotal information, as the public data does not list Hondurans among the 16 largest migrant groups.

[8]Interviews, Tegucigalpa, July–August 2014.

[9]Much of this violence disproportionately affects young men. Of recorded homicide victims, 93.1 % were male (UNODC 2011) and according to the Small Arms Survey (2014), "when calculating only for the male population aged 15–49, 33 % of all deaths were attributable to violence".

maras but is also used as a way of furthering the groups' economic activities, such as extortion, kidnapping and involvement in the sale and transport of drugs, mainly marijuana, cocaine and crack. Although the *maras* originated in the US and were imported to the region through deportation, their violence has been exacerbated by local policies, such as in response to iron fist (*mano dura*, literally 'hard hand') approaches of the state authorities and increased involvement in the narcotics trade (Cruz 2014). Although gang membership and exposure to lethal violence is predominantly young and male, women and girls play a larger role in the Central American *maras* than usual for urban gangs and are also exposed to high levels of physical abuse and SGBV, as well as targeting by anti-gang death squads along with their male colleagues (Interpeace 2013).

Other forms of violence in urban areas include targeted killings of women (especially of young women employed in the *maquilas* of the textile industry), substance abuse, the so-called "social cleansing" (*limpieza social*) of suspected *maras* and lower-class teens by death squads, domestic violence and SGBV (on various forms of violence see e.g. Pine 2008; Small Arms Survey 2014). Rural violence tends to take on other forms and is often related to either land issues or the control of trans-shipment routes of drugs (Gillard 2013; International Crisis Group 2014; La Tribuna 2013). Domestic violence and SGBV are also major concerns in rural areas. The various forms of physical violence which Hondurans are exposed to is thus linked closely to gender, age, class and location. In recent years, the indirect impacts of climate change look to be increasing the likelihood of Hondurans being exposed to these forms of violence.

5.5.3 Coffee

Coffee is the main export product of Honduras, which is the region's largest exporter. Central American coffee crops are, however, increasingly being affected by coffee leaf rust fungus caused by the fungal plant pathogen *Hemileia vasatrix*, known locally as *roya,* the spread of which is linked to climate change (ICO 2013; IICA 2013; Oxfam 2014). The impacts of this on local economies in El Salvador, Guatemala, Honduras and Nicaragua are potentially immense, with regional direct job losses estimated to reach 300,000–400,000 over the next few years, which, assuming an average household size of 5.25 persons (Tucker 2008), would directly impact 1,575,000–2,250,000 people. Based on initial research by Fairtrade Finland, the impacts are not spread evenly: those most affected tend to be the poorest farmers who have not been able to care for their shrubs as effectively as wealthier landowners have.[10] Given the substantial amount of time and money that needs to

[10]Personal communication with Janne Sivonen, Executive Director, Fairtrade Finland, May 2014 and Professor Catherine Tucker, Indiana University, June 2014. This seems to be corroborated by initial findings from other research in the area.

be invested in coffee, farmers are also often reluctant to give up on coffee farming once they have invested in it.

It is important to note, however, that the structure of the coffee industry and patterns of land ownership differ greatly between the various countries in the region (Paige 1998). In El Salvador and Costa Rica, for example, the sector continues to be mainly dominated by large estates (*fincas*) owned by few powerful families, while in Nicaragua the Sandinista Revolution led to more small-holders and cooperatives in the coffee sector. In Honduras, coffee plantations tend to be much smaller than in its neighbouring countries and ownership is spread over a larger number of small- to medium-scale landowners (Reichman 2011; Tucker 2008). These structural differences place the coffee farmers and labourers in different situations of vulnerability to the impacts of *roya*: larger plantations will have more capital available to adapt than smallholders, but smallholders may still be economically and socially less vulnerable than the hired labourers working on larger *fincas*.

5.5.4 Coping with Climate Change, Violence and Migration

A first, key step in designing gender-responsive peacebuilding programming around the complex societal knock-on effects of climate change-induced *roya* is understanding to what degree different households are affected and what the gendered coping strategies will be. In the initial phase, households will likely aim to reduce their expenses and perhaps take on loans, processes that will inevitably involve decisions about prioritising certain needs over others. Given difficulties of finding alternate employment in rural areas, it can also be expected that the loss of coffee income may well contribute to increased migration both into the USA and Central American urban centres. Many of the urban centres are, however, already are struggling with the equitable provision of services to citizens, limited employment opportunities and are also, as mentioned previously, heavily impacted by different forms of gendered violence, including extremely high rates of homicide. Rural-urban migration may be further exacerbated by severe droughts that hit Honduras in 2014, but reliable data on population movements related to this were not available at the time of writing (IFRC 2014).

Although gang-related violence is a major factor in Central American urban spaces, its impacts are often sensationalized and, based on initial reviews of detailed, long-term ethnographic studies of the *maras*, it is unlikely that rural migrants would directly join gangs due to age and the sub-cultural dynamics of gang socialisation (Brenneman 2012; Ward 2012). There does, however, seem to be a possible 'second-generation' impact of the migration of parents and subsequent (and at times abusive) foster care of children left behind and the latter's propensity to join urban gangs. Thus, it might well be the daughters and sons of migrants, rather than migrants themselves, who join the 'surrogate families' provided by the *maras* (Brenneman 2012; Ward 2012; on the 'missing generation of parents', see also Reichman 2013). Urban violence and fear thereof is also increasingly becoming a 'push factor' for Hondurans to emigrate (Reichman 2011; UNHCR 2014).

In the rural areas, it is likely that the ones opting to migrate would tend to be the ones who the rural areas can least afford to lose, i.e. the younger and more dynamic members of society. Whether men and women will migrate on their own or as families remains to be seen and will be analysed as part of this study. In the past, it has been mostly men or women migrating on their own, leaving their children to be cared for by the extended family, but increasingly, based on anecdotal evidence, whole families seem to be migrating and/or children are migrating to join their parents abroad.

While the gendered impacts of *roya* for different men and women in Honduras remain to be studied in detail, a possible trend that is already emerging is that rural, lesser educated, young men's access to traditional coping strategies seems to be increasingly reduced. With the legal economic opportunities in the rural economy threatened, migration to urban areas and to the US have been long-established adaptation strategies for men (Reichman 2011).[11] However, urban migration is becoming an increasingly unsavoury option given the rising rates of violence (and perceptions thereof), and jobs in the urban economy tend to be relatively feminised, both in the *maquilas* of the textile industry and in the service sector (Pine 2008). As in other Central American countries, the one sector that is open for men in urban areas with low educational backgrounds, especially rural-to-urban migrants, is the increasingly precarious job of being a security guard (Dickins de Girón 2011). With narrow options available in the cities, migration north has been a more alluring option, but this route might become increasingly unavailable as the US tightens its border security and presses for its southern neighbours to do the same.

In terms of peacebuilding, understood in the broad sense of helping communities find non-violent ways of dealing with conflict, what we aim to do with this project is to identify local networks in the form of state administration (especially at the municipal/*barrio* level) as well as civil society organisations (including women's and men's organisations, anti-violence campaigners, churches and faith-based organisations and more social work-oriented groups), which work both directly and indirectly on reducing various forms of violence and increasing social inclusivity.

Whether or not these networks, mechanisms, institutions and individuals are resilient enough to cope with a potentially massive, sudden influx of new, impoverished and poorly skilled individuals or families (in the urban case), and the creation of sustainable livelihoods or the temporary or permanent outmigration of these individuals and families (in the case of rural communities) is currently not known.

What is, however, evident is that communities with functioning mechanisms and cultures of local-level governance, conflict settlement and agreed-upon regulatory

[11]One possible new rural 'growth industry' that might provide new, but limited economic opportunities is the increasing trans-shipment of narcotics through Honduras (International Crisis Group 2014).

mechanisms look to be better placed to cope with the social and economic fallout of environmental stress than those which do not have these characteristics (Tucker 2008). This might place certain rural communities with higher levels of social cohesion at an advantage, assuming the impacts of the crisis do not overburden these capacities. The key questions will be how these can be supported and strengthened in those areas where they exist, how they can be established where they do not, and how they can be made inclusive in terms of gender, class, age and other factors.

Devising programming together with local partners and beneficiaries will require building an understanding of what a loss of income from coffee production means for different women and men; what coping strategies are available to them; what kind of forms of violence (e.g. domestic, interpersonal, criminal or self-inflicted violence such as substance abuse or suicide) increased socio-economic stress might exacerbate and for whom; what vulnerabilities boys, girls, men, women and sexual and gender minorities of different ages and classes might face; and what these dynamics mean for gendered societal power relations in Honduras. Building on an in-depth understanding of these changing dynamics, local-level peacebuilding efforts would then be supported, in turn increasing local resilience, gender equality and social inclusion and, ideally, sustainable coping strategies for dealing with the impacts of climate change.

5.6 Conclusion

While the need to 'somehow' bring together gender, peacebuilding and climate change adaptation has been recognised for at least over a decade and a half, what exactly that might mean in practice has remained unclear. Some responses have been overly simplistic, while other approaches have been so entangled in local complexities they have become unusable.

What we are attempting with our project in Honduras with our partners—and other programming on gender, climate change and peacebuilding—is to find a middle way. We want to use a gender relational lens to better understand power dynamics and identities which shape people's vulnerabilities and modes of resilience open to them, their abilities of having agency and control, and the particular needs and concerns they may face. While this is at first sight a complex task with a vast range of variables, spending more time at the outset to understanding these will ideally make our work simpler and more effective in later stages. It also requires listening to the local actors and beneficiaries, as the men, women, boys and girls of the coffee-growing areas of Honduras are making gendered analyses of their situation and various options on a daily basis, weighing up what the dangers and opportunities of various strategies mean for them, their families and communities. Based on an analysis of these factors and dynamics, we aim to develop gender-responsive peacebuilding interventions together with the local communities, in order to enhance their capacities to deal with the multiple gendered challenges posed by climate change.

References

Brenneman, Robert, 2012: *Homies + Hermanos: God and Gangsters in Central America* (Oxford: Oxford University Press).

CCSPJP (Consejo Ciudadano para la Seguridad Pública y la Justicia Penal), 2012: *San Pedro Sula, la ciudad más violenta del mundo; Juárez, la segunda* (Mexico, DF: CCSPJP).

Cohn, Carol (Ed.), 2012: *Women and Wars: Contested Histories, Uncertain Futures* (London: Polity Press).

Cruz, José Miguel, 2014: "Maras and the Politics of Violence in El Salvador", in: Hazen, Jennifer; Rodgers, Dennis (Eds.): *Global Gangs: Street Violence across the World* (Minneapolis: University of Minnesota Press): 123–144.

Dickins de Girón, Avery, 2011: "The Security Guard Industry in Guatemala: Rural Communities and Urban Violence", in: O'Neill, Lewis; Thomas, Kevin; Thomas, Kedron (Eds.): *Securing the City: Neoliberalism, Space, and Insecurity in Postwar Guatemala* (Durham: Duke University Press): 103–126.

Dolan, Chris 2001: *Collapsing Masculinities and Weak States—A Case Study of Northern Uganda* (unpublished).

Dolan, Chris, 2002: "Collapsing Masculinities and Weak States—A Case Study of Northern Uganda", in: Cleaver, Frances (Ed.): *Masculinities Matter!—Men, Gender and Development* (London: Zed Books): 57–83.

El-Bushra, Judy, 2012: *Gender in Peacebuilding: Taking Stock* (London: International Alert); at: http://www.international-alert.org/sites/default/files/publications/201210GenderPeacebuilding-EN.pdf (24 April, 2015).

El-Bushra, Judy; Naujoks, Jana; Myrttinen, Henri, 2014: *Renegotiating The 'Ideal' Society: Gender in Peacebuilding in Uganda* (London: International Alert); at: http://www.international-alert.org/resources/publications/renegotiating-ideal-society (24 April, 2015).

Gillard, Michael, 2013: "British Gas Risks Fuelling Dirty War in Killing fields of HONDURAS", in: *The Sunday Times*, December 8, 2013.

Gutmann, Matthew, 2006: *The Meanings of Macho: Being a Man in Mexico City*, Tenth Anniversary Edition (Berkeley: University of California Press).

Herrera, Natalia; Porch, Douglas, 2008: "'Like Going to a Fiesta'—The Role of Female Fighters in Colombia's FARC-EP", in: *Small Wars & Insurgencies*, 19,4: 609–634.

Hirsch, Sarah, 2010: "Migration and Remittances—The Case of Honduras", in: *Rural 21 Focus* (Tegucigalpa: GTZ Honduras).

ICO (International Coffee Organisation), 2013: *Report on the Outbreak of Coffee Leaf Rust in Central America and Action Plan to combat the pest* (London: International Coffee Organisation); at: http://dev.ico.org/documents/cy2012-13/ed-2157e-report-clr.pdf (24 April 2015).

IFRC (International Federation of Red Cross and Red Crescent Societies), 2014: *Emergency Appeal Honduras: Drought, Emergency Appeal n° MDRHN008* (Geneva: IFRC, 13 October 2014).

IICA (Instituto Interamericano de Cooperación para la Agricultura), 2013: *La crisis del café en Mesoamérica – Causas y respuesta apropiadas* (Washington DC: IICA); at: http://www.iica.int/Esp/prensa/BoletinRoya/2013/N01/Roya-MA.pdf (24 April 2015).

INE (Instituto Nacional de Estadística) 2014: *Population Figures at 1 January 2014—Migration Statistics 2013* (Madrid: INE).

International Alert, 2010a: *Programming Framework for International Alert – Design, Monitoring and Evaluation* (London: International Alert).
International Alert, 2010b: *Changing Fortunes: Women's Economic Opportunities in Post-War Northern Uganda* (London: International Alert).
International Crisis Group, 2014: *Corridor of Violence: The Guatemala-Honduras Border*, Latin America Report N°52 (Brussels: International Crisis Group).
Interpeace, 2013: *Violentas y violentadas – Relaciones de género en las maras Salvatrucha y Barrio 18 del triángulo norte de Centroamérica* (Guatemala City: Interpeace Regional Office for Latin America).
IPCC (Intergovernmental Panel on Climate Change), 2001: *Climate Change 2001: Impacts, Adaptation and Vulnerability, Contributions of the Working Group II to the Third Assessment Report of the Intergovernmental Panel on Climate Change* (Cambridge: IPCC).
Jauhola, Marjaana, 2013: *Post-Tsunami Reconstruction in Indonesia: Negotiating Normativity through Gender Mainstreaming Initiatives in Aceh* (New York: Routledge).
Knight, Kyle 2014: "Lost in the Chaos—LGBTI People in Emergencies", IRIN News, August 14, 2014; at: http://www.irinnews.org/report/100489/lost-in-the-chaos-lgbti-people-in-emergencies (24 April, 2015).
La Tribuna, 2013: "Bandas Criminales Tienen a San Luis Como el Municipio Mas Violento de Honduras", *La Tribuna*, 4 March 2013.
Myrttinen, Henri; El-Bushra, Judy; Naujoks, Jana, 2014: *Re-Thinking Gender in Peacebuilding* (London: International Alert); at: http://international-alert.org/sites/default/files/Gender_RethinkingGenderPeacebuilding_EN_2014.pdf (25 April 2015).
Neumayer, Eric; Plümper, Thomas, 2007: "The Gendered Nature of Natural Disasters: The Impact of Catastrophic Events on the Gender Gap in Life Expectancy, 1981–2002", in: *Annals of the Association of American Geographers*, 97,3: 551–66.
Ormhaug, Christin Marsh; Meier, Patrick; Hernes, Helga, 2009: *Armed Conflict Deaths Disaggregated by Gender*, PRIO Paper (Oslo: PRIO).
Oxfam, 2014: *Afectación de la Roya en los medios de vida de Productoras y Productores Jornaleros del Café en Honduras* (Tegucigalpa: Oxfam); at: https://www.oxfam.org/sites/; www.oxfam.org/files/estudio_roya_hnd.pdf (24 April, 2015).
Paige, Jeffery M., 1998: *Coffee and Power: Revolution and the Rise of Democracy in Central America* (Cambridge: Harvard University Press).
Peters, Katie; Vivekananda, Janani, 2014: *Topic Guide: Conflict, Climate and Environment* (London: Overseas Development Institute/International Alert).
Pine, Adrienne, 2008: *Working Hard, Drinking Hard—On Violence and Survival in Honduras* (Oakland: University of California Press).
Reichman, Daniel, 2011: *The Broken Village—Coffee, Migration, and Globalization in Honduras* (Ithaca: Cornell University Press).
Reichman, Daniel, 2013: "Honduras: The Perils of Remittance Dependence and Clandestine Migration", in: *The Online Journal of the Migration Policy Institute*; at: http://www.migrationpolicy.org/article/honduras-perils-remittance-dependence-and-clandestine-migration (25 April 2015).
Roy, Anirban, 2008: *Prachanda: The Unknown Revolutionary* (Kathmandu: Mandala Book Point).
Small Arms Survey, 2014: *Firearms and Violence in Honduras*, Research Note 39 (Geneva: Small Arms Survey).
Tucker, Catherine, 2008: *Changing Forests—Collective Action, Common Property, and Coffee in Honduras* (Heidelberg: Springer).
Turner, Simon, 1999: *Angry Young Men in Camps: Gender, Age and Class Relations Among Burundian Refugees in Tanzania*, Working Paper No. 9 (Roskilde: Roskilde University–UNHCR).
UNHCR, 2014: *Children on the Run—Unaccompanied Children Leaving Central America and Mexico and the Need for International Protection* (Washington DC: The United Nations High

Commissioner for Refugees Regional Office (UNHCR) for the United States and the Caribbean).

United Nations, 2013: *Women and Natural Resources: Unlocking the Peacebuilding Potential* (New York–Geneva: UNEP, UN Women, UNDP–UN PBSO).

UNODC (United Nations Office on Drugs and Crime), 2011: *2011 Global Study on Homicide—Trends, Contexts, Data* (Vienna: UNODC).

Urdal, Henrik, 2010: "Mythbusting: Are Women the Main Victims of Armed Conflict?", in: *PRIO Gender, Peace and Security Update July–August 2010 Issue* (Oslo: PRIO): 1–2.

Vigh, Henrik, 2006: *Navigating Terrains of War—Youth and Soldiering in Guinea-Bissau* (Oxford: Berghahn Books).

Vivekananda, Janani; Schilling, Janpeter; Smith, Dan, 2014: "Climate Resilience in Fragile and Conflict-Affected Societies: Concepts and Approaches", in: *Development in Practice*, 24,4: 487–501.

Ward, Thomas, 2012: *Gangsters Without Borders: An Ethnography of a Salvadoran Street Gang* (Oxford: Oxford University Press).

WOLA (Washington Office on Latin America), 2014: *Mexico's Other Border: Security, Migration, and the Humanitarian Crisis at the line with Central America* (Washington DC: WOLA); at: http://www.wola.org/publications/mexicos_other_border (23 April 2015).

World Bank, 2011a: *World Development Report: Conflict, Security, and Development* (Washington DC: World Bank).

World Bank, 2011b: *Gini index (World Bank Estimate)* (Washington DC: World Bank); at: http://data.worldbank.org/indicator/SI.POV.GINI/ (24 April 2015).

Wright, Hannah, 2014: *Masculinities, Conflict and Peacebuilding: Perspectives on Men Through a Gender Lens* (London: Saferworld).

Yami, Hisila, 2007: *People's War and Women's Liberation in Nepal* (Kathmandu: Janadhwani Publications).

Commissioner for Refugees, Regional Office (UNHCR) for the United States and the Caribbean.

Chapter 6
The Water, Energy, Food and Biodiversity Nexus: New Security Issues in the Case of Mexico

Úrsula Oswald Spring

Abstract This chapter analyses the security nexus between *water, energy, food and biodiversity* (WEF&B). The research question is, how could the nexus between WEF&B security be improved in a country with high environmental and social vulnerability, and which is seriously affected by climate change and organized crime? After a short conceptual review of WEF&B security, the dominant nexus is explored for Mexico, addressing first the feedbacks between water and biodiversity, and later changes in land use, food production and social vulnerability. Mexico is an oil-exporting country and has the fourth most important reserve of shale gas in the world. It has extensive drylands where 77 % of the population lives. These produce 87 % of the GDP but receive only 31 % of the water that falls as rainwater; the environment and the aquifers are thus overexploited. Furthermore, a neo-liberal free trade policy has allowed highly subsidized food imports, as well as rural–urban and international migration of peasants. Finally, extreme events influenced by climate change, such as hurricanes and droughts, have had a negative impact on human lives and on the economy. In addition, organized crime controls a part of the trade in migrants, drugs, and arms, as well as timber. A weak legal system has fostered small-scale crime, and this has increased public insecurity. As well as this, fracking activities in water-scarce regions are impacting on deep aquifers and limiting processes of adaptation to climate change in desert regions. The nexus between scarce water, overexploited aquifers, deforested areas, disasters, high food prices, weak rural government support, high energy prices and fragile governance is increasing poverty and the migration of farmers on rain-fed lands, as well as creating the risk of social instability in urban areas.

Prof. Dr. Úrsula Oswald Spring, full-time Professor/Researcher at the National University of Mexico (UNAM) in the Regional Multidisciplinary Research Center (CRIM); Email: uoswald@gmail.com.

This research was financed by PAPIIT–DGAPA Project IN300213 of the National Autonomous University of Mexico, entitled "Integrated water management of a river basin affected by climate change: risks, adaptation and resilience". I thank Anahí Bustamante for her participation in the risk survey, Hans Günter Brauch for academic discussions, the anonymous peer-reviewers for their critical input. I am also grateful to Mike Headon for his careful English revision.

H.G. Brauch et al. (eds.), *Addressing Global Environmental Challenges from a Peace Ecology Perspective*, The Anthropocene: Politik—Economics—Society—Science 4, DOI 10.1007/978-3-319-30990-3_6

Keywords Nexuses between water, energy, food and biodiversity security (WEF&B) · Climate change · Disasters · Food insecurity · Adaptation · Organized crime · Social instability · Human security

6.1 Introduction

This chapter reviews the emergence of the policy debate on the nexus between energy, water and food security, influenced by conceptual inputs from the *World Economic Forum* (WEF 2013). In its initial paper the WEF took a security approach (with reference to water, energy, and food security) related to the *business-as-usual* policy of the global oligarchy.[1] Their Hobbesian approach is preparing the political arena to deal militarily with upcoming conflicts resulting from environmental and social disasters (Melillo et al. 2014) in order to maintain the existing economic order and power structure.

In contrast, this chapter addresses the nexus between *water security* (WS), *energy security* (ENS), *food security* (FS) and *biodiversity security* (BS) from an *environmental* (ES) and *human security* (HS) perspective. This HS approach focuses on the pillars of *freedom from want* (CHS 2003) and *freedom from hazard impacts* (Bogardi/Brauch 2005). In the case of the drug war it also includes a third pillar, *freedom from fear* (UNDP 1994). This text analyses the impacts for Mexico of a *business-as-usual* policy on the *water, energy, food, and biodiversity* (WEF&B) security nexus, which has increased poverty in large parts of the population and produced environmental destruction. Mexico has both Neartic and Neotropical ecosystems with an exceptional level of biodiversity. It is also greatly affected by climate variability because of its location in the tropics, its extensive drylands, and its 11,000 km of coastlines between two warming-up oceans. It depends for its agriculture and its water supply on the summer monsoon, and this is increasingly becoming more variable. It is exposed to hurricanes, flash floods, and droughts, and it also faces geophysical hazards (earthquakes, tsunamis, and volcanic eruptions; Fig. 6.1; MunichRe 2008). Its topography is predominantly mountainous and soils are exposed to erosion and desertification because of massive deforestation and inadequate agricultural management.

All these natural and anthropogenic phenomena have increased dual (environmental and social) vulnerability (Bohle 2002; Oswald Spring 2013), which is aggravated by the present neo-liberal policy of privatization of public services (education, health, water, sanitation, etc.), reduction in salaries, targeted alleviation of poverty, high interest rates, massively subsidized food imports, a low level of

[1]The world economy and an important element of policy formulation is ruled by a number of multinational corporations, which control the financial (stock market, banks, and tax havens), productive, commercial, and entertainment institutions. This globally controlled and highly interconnected financial, productive, trade, military and political system is defined in this article as global oligarchy.

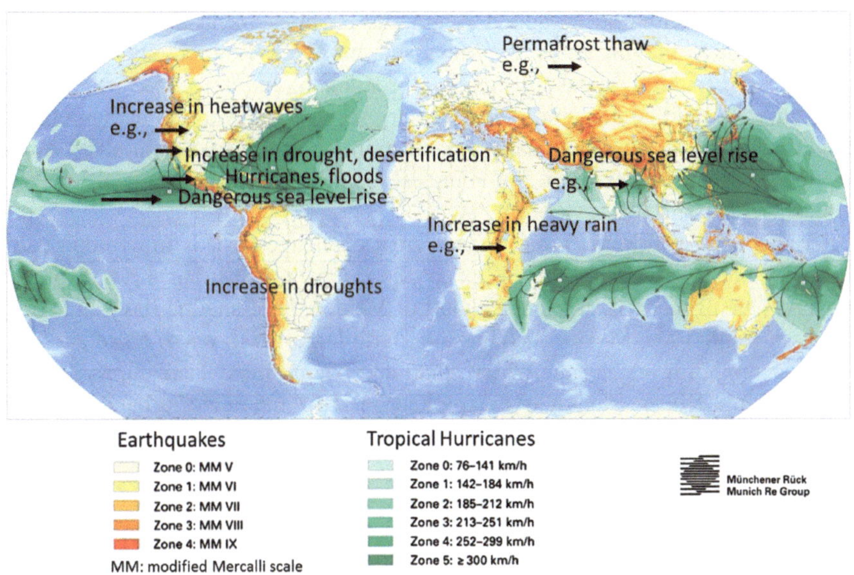

Fig. 6.1 World map of natural hazards. *Source* MunichRe (2008)

investment in science and technology, and a lack of credits for small enterprises. In 1982 Mexico was the first country to agree to a structural adjustment programme with the *International Monetary Fund* (IMF). This was later fully implemented with a massive level of privatization by a dependent national government in cooperation with the national and international oligarchy (Delgado/Gutiérrez 2007). In 2014, more than eight million young people had no jobs and no opportunity to study. Organized crime took advantage of this situation by involving them in illegal activities. It widened its activities from drug-trafficking to kidnapping, extortion, pornography, and trafficking of humans, organs, arms, timber, water, oil, minerals, endangered species, and archaeological artefacts. These criminal activities were concentrated in environmentally diverse regions and on the trade routes to the US, and affected highly vulnerable regions such as Guerrero, Morelos, Michoacán[2] (De la Torre González et al. 2011), Chihuahua, Sonora, and Tamaulipas, the border states with the USA. The result has been an increase in violence and a higher level of dual vulnerability in a country where the laws are weakly enforced. This has resulted in public insecurity, high homicide rates and a deterioration in natural and social conditions.

The natural climate conditions, the biodiversity, and the surface and groundwater present an important potential for human development, and this includes the drylands. Indigenous societies have domesticated and adapted maize, beans, squash, and other food products of the tropical humid ecosystem and taken them from sea

[2]Together with the state of Mexico, this state sends one-fifth of the water supply to the Metropolitan Valley of Mexico City through the Cutzamala system (Morales/Rodríguez 2011).

level up to the semi-arid mountain areas. These achievements contributed to the emergence of two civilizations: the Maya and the Aztec. Both empires intensively exploited the fragile tropical soil for food production, and developed irrigation so that harvests could be gathered during the dry season thanks to the benign climate conditions. This also favoured early urbanization. The overuse of natural resources, together with internal conflict and conquest, contributed to the downfall of these sophisticated civilizations (Coe 1999).

Contemporary Mexican society faces great challenges if it is to overcome this high level of dual vulnerability. Mexico has reserves of oil, but the unexploited resources are primarily located in the Gulf of Mexico, necessitating risky and costly deep-sea exploration. Mexico also has the fourth largest reserves of shale gas in its water-scarce north (e.g. in the Chihuahuan and Nuevo León-Tamaulipas deserts), where the aquifers are already overexploited. In 2014, the rate of urbanization was 78.3 %. Sixty-four per cent of the Mexican population works in the informal sector, with no social protection or regular income. Coneval (2013; responsible for poverty analysis) stated that 74.2 % of Mexicans face at least one type of social deprivation and only 19.9 % do not face any. This high level of social vulnerability is a result of complex internal and external pressures, and the analysis of the WEF&B nexus is crucial to understand the outcome of dual vulnerability.

After presenting the research questions and hypothesis, this chapter conceptualizes human security in relation to *freedom from want* and *freedom from hazard impacts*, explains the concept of dual vulnerability, and presents an interrelated scheme between WEF&B nexuses. With this conceptual background, the chapter explores the positive and negative nexuses between WEF&B in Mexico. A process of the securitization of dual vulnerability is put forward as a political tool to promote emergent actions that will overcome both vulnerabilities. After exploring the hidden agenda of the hard security nexuses of the global oligarchy, in its conclusions the chapter proposes a paradigm shift from *business-as-usual* pursued by this global oligarchy[3] to a sustainable transition that recognizes that there is only one planet Earth with conditions for life and joy, and that this oligarchy also has to live here. This shift from military to human and environmental security aims at a reduction in dual vulnerability and a reduction in military spending, and opens up the path from *business-as-usual* towards a human and environmentally friendly future. Mexico will be the subject of the empirical discussion, and the chapter will examine to what extent this assessment may contribute to reframing policy goals in this country.

[3]The global oligarchy are mentioned regularly in Forbes as the richest people on Earth, and UNDP (1994) showed that eighty-five of the richest people own the same wealth as 3.5 billion poor people. This means that wealth is highly concentrated and that these oligarchs, whatever country they live in (mostly the US) have investments in multinational businesses: production, trade, services, finance, and the military and arms industries.

6.1.1 Research Questions and Hypothesis

What new scientific insights might overcome the narrow understanding of security in the nexuses between water, energy, food and biodiversity security, and to what an extent might such an approach contribute to reframing the goals of policies that deal with the linkages between these nexuses? How can an approach centred on human security suggest both top-down and bottom-up activities to reduce this dual vulnerability?

The hypothesis of this chapter is that the nexus between water, food and energy recognized by the WEF is a new way of hiding the oligopolistic expansion of global capital, using a narrow concept of military security to achieve the oligarchy's goals at the cost of the majority of human beings and of the environment, while a human security approach may offer a transition to sustainability.

6.1.2 Conceptualization of Human Security, of the Securitization of Dual Vulnerability, and of the Nexuses Between Water, Energy, Food, and Biodiversity Security

Since the World Economic Forum of 2012, a number of governments and scholars have begun a conference marathon addressing different security nexuses (water, energy, and food security), but where the term 'security' has remained mostly undefined. Most governments have retained their state-centred focus on security, while some academics have preferred a human security approach. This recent securitization of multiple nexuses has increased the sense of urgency but the political will has been lacking to initiate extraordinary measures to reduce the emission of *greenhouse gases* (GHG) and to put a halt to environmental deterioration, especially biodiversity loss. The nexus proposals may in fact be preparing a hidden agenda for coming conflicts related to climate change and resource scarcity. This debate also distracts attention from the launching of international, legally binding commitments, and thus indirectly justifies the dominant military and political understanding of security. It limits concessions on international agreements, maintains the destruction of the environment, and increases social vulnerability, all in the name of national or international security.

Using a postmodern understanding of security (Brauch/Scheffran 2012), deterritorialization is essential to the debate on human security, as it is for the sectoral concepts of water, energy, food, and biodiversity security. The *United Nations Development Programme* (UNDP 1994) developed the *human security* (HS) concept and shifted the reference object to humans and risks (Beck 2009, 2011) to their survival. Ogata and Sen (CHS 2003) reviewed the structural factors of poverty, inequality, conflicts and social anomie in the context of globalization and pointed to *freedom from want* to highlight the precarious conditions of more than half of the world's population. Bogardi/Brauch (2005) and Brauch (2005), by referring to *freedom from hazard impacts,* focused on social vulnerability, and the necessity of building resilience so that people affected by climate and *global environmental change* (GEC) can adapt.[4] Both require efforts to counter corruption and a reinforcement of the state of law together with respect for human rights so that people can live in *freedom with dignity* (Annan 2005).

Human and gender security are both understood in this chapter as deepened security concepts comprising the individual, the community up to the national level, and the global society (Brauch et al. 2008). Focusing on human security, the reference object has shifted from the state and sovereignty—which is how the WEF has traditionally understood security—to humankind. The values at risk are the survival of people and their cultural identity. The sources of threat are globalization, extreme events, poverty, lack of a state of law, and corruption among governments and multinational enterprises. In the case of gender security, the reference objects are women, children, the indigenous, the elderly, minorities, and all those who lack power. The values at risk are gender relations, equity, identity, and social representation, while the source of threats is the patriarchal system, which has emerged over thousands of years. This violent system is today represented by oligarchic elites who impose an inhuman globalization, authoritarian states, discriminative institutions, and hierarchical churches that have ideological control over the faith of the people. Wealth has been concentrated in a tiny oligarchy which has often favoured authoritarian and intolerant institutions. Monopolized mass media have fostered a consumerist and wasteful culture. From the perspective of human and gender security, nature is also becoming a threat to humans through extreme events that often turn into disasters because of the lack of preventive behaviour and adaption. At the global scale, a lack of consciousness and the blocking of a climate change agreement among dominant governments are key mechanisms to maintaining the status quo and thus increasing dual vulnerability for most human beings around the world.

The concepts of *Environmental* (ES), societal and economic security were introduced by the Copenhagen School as part of the widening of the security process (Buzan et al. 1998). In ES the reference objects are rural and urban ecosystems, water and food. The value at risk is sustainability and the sources of threats are humankind and nature. For the first time humans, with their intensive use of fossil energy and their wasteful consumerism, are creating threats to their own survival, but at the same time they are the victims of *global environmental change* (GEC) caused by extreme events (IPCC 2012, 2013, 2014). From a constructivist

---

[4]GEC is more than climate change. It includes in the natural system changes in the chemical-physical composition of the atmosphere, the soil, the biota, water, and subsoil. In the human system it refers to population growth, transformation in the rural and urban system, and changes in the productive processes (Brauch et al. 2008, 2009). Both the natural and the human system interact and produce negative feedbacks e.g. the emissions of greenhouse gases increase the threat of hazards, which may turn into disasters when people lack early warning mechanisms, preventive evacuation procedures, adaptation and resilience.

perspective, Wæver (2008: 582) proposed a process of securitization consisting of three components:

(a) the *securitizing actor*, i.e. an entity that makes the securitizing move;
(b) the *referent object* that is being threatened, and the values to be protected;
(c) the *audience*, which is the target of the securitization act.

The audience needs to accept the securitized reference object as a crucial security threat. The securitization move is considered successful only when the audience is convinced and accepts measures and constraints. From a human security perspective, hardly any substantial changes have taken place in the international arena to achieve the stated goals; goals were adopted and action towards them was only partially implemented (see Millennium Development Goals).

Since the 1970s different sectoral security concepts have emerged. The *International Energy Agency* (IEA) has defined energy security (ENS) as:

> the uninterrupted availability of energy sources at an affordable price.… With particular emphasis on oil security, the IEA was created to establish effective mechanisms for the implementation of policies across a broad spectrum of energy issues: mechanisms that were workable and reliable, and could be implemented on a co-operative basis.[5]

The concept of *food security* was introduced in 1974 by the FAO (1983), and changes in its definition reflect ongoing policy debates. The concept has been contested by the farmer's movement *Via Campesina*, who introduced the concept of *food sovereignty* (Oswald Spring 2009b). *Water security* (WS) was introduced and defined during the Ministerial World Water Forum at The Hague (2000). From a human security perspective, in this text water security is oriented at a widened and deepened environmental, societal, economic, food and ecological security approach

> acknowledging water's life-giving characteristics. It includes the *avoiding* of difficulties in terms of pollution and silting linked to water's lift-up/carry away functions and its mobility, to be achieved by water pollution abatement and soil protection. *Foreseeing* unavoidable conflicts and difficulties linked to climatic variability (droughts, floods), and to water's multiple functions and mobility reflected in non-negotiable natural processes in the landscape (Oswald Spring/Brauch 2009a: 180).

The concept of *biodiversity security* (BS) has so far not been conceptualized. For the FAO (2010), "biodiversity is essential for food security and nutrition", and global food production depends on "a vital web of biodiversity within the ecosystems". The FAO further claims that "with the erosion of biodiversity, humankind loses the potential to adapt ecosystems to new challenges such as population growth and climate change".[6] BS supports and conserves *ecosystem services* (ESS), which provide, support, and regulate natural processes, and create non-material cultural benefits. The reference object focuses on natural ecosystems, animals and plants. The value at risk is sustainability and the sources of threats

[5]See at: http://www.iea.org/topics/energysecurity/.

[6]FAO: 'Biodiversity', at: http://www.fao.org/biodiversity/en/.

relate to humankind and nature, including extreme events. A stable biodiversity security maintains natural and agricultural production, soil fertility and the nutrient cycle, together with all ESS. BS is thus an essential component of food security, nutrition and human health.

The sum of human, gender, and environmental security—including its sectoral components of water, energy, food, and biodiversity security—has been referred to as a HUGE security (Oswald Spring 2009a). HUGE includes a widened, deepened, and sectoralized understanding of the security concept. The reference objects are both the environmental and human systems; the values at risk are sustainability, livelihood, equality and equity, and the sources of threats are human activities controlled by the oligarchy.

Dual vulnerability interrelates environmental and social factors in a context of globalization (Fig. 6.2). On the environmental side, pollution, the loss of natural soil fertility, deforestation, the destruction of crucial ecosystem services, and climate variability have aggravated hazards through extreme events (floods, flash floods) and lack of water (drought). Social vulnerability has also been increased through chaotic urbanization, change in land use from forests to agriculture and human settlements, scarce resources, and failed harvests. Massive imports of basic food and weak protection for peasants on rain-fed land have forced entire

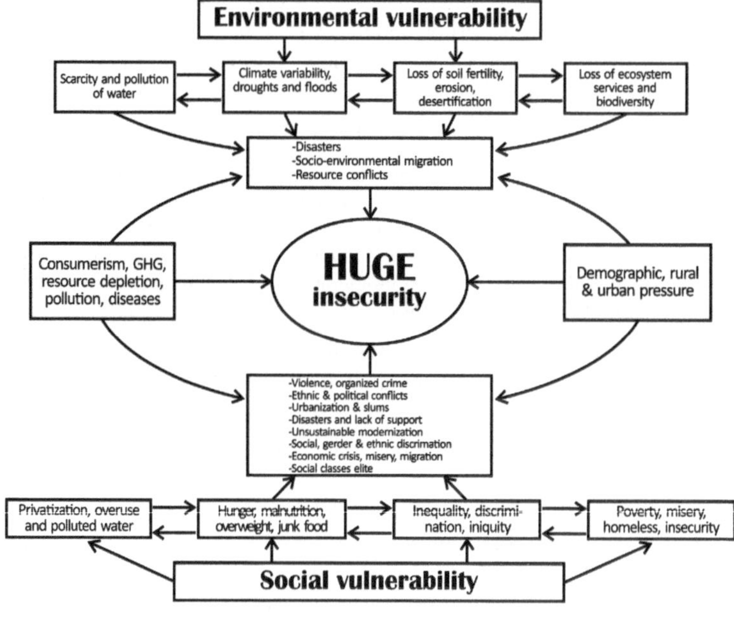

Fig. 6.2 Dual vulnerability: environmental and social vulnerability. *Source* Developed by Oswald Spring (2013), inspired by Bohle (2002)

communities to migrate to cities or abroad. In cities, precarious jobs without social security, long transportation times, and public insecurity are presenting human life with a high level of social vulnerability. There is no doubt that households headed by women with a low level of education (24.6 % of families in Mexico) are in the most precarious situation.

These brief conceptual considerations allow the focus to be shifted from a state-centred hard security approach to a human, gender and environmental-centred security approach. The next section discusses the nexus of WEF&B, first in general terms and then specifically for Mexico.

6.2 Nexus Between Energy, Food, Water and Biodiversity Security

Figure 6.3 addresses the crucial nexus between water and energy; water and biodiversity; water and food; energy and food; food and biodiversity; and energy and biodiversity in the current political arena of neo-liberalism. Key national figures act as part of a transnational oligarchy. Their productive and service systems are integrated within multinational enterprises and interlinked with global financial flows, the international homogenization of culture, fashion, military expenditure, and the arms trade. Their interaction maintains *business-as-usual* and concentrates wealth in the hands of a small oligarchy. They exercise pressure on governments through lobbying, bribes, and support for political campaigns. Through television, films, and social networks they influence global society, and through propaganda and fashion they promote a consumerist behaviour that benefits their business. International organizations—the International Monetary Fund, the World Bank, and the World Trade Organization—support this global neo-liberal model.

In the international political arena, no powerful institution exists to negotiate global governance. In the security area, the five permanent members of the Security Council of the United Nations (United States, China, Russia, United Kingdom and France) may veto decisions that challenge their geopolitical interests. This global arena and its actors promote a policy of *business-as-usual,* causing human insecurity, especially that of the most vulnerable (UNEP 2014). The multiple nexuses between water, food, energy, and biodiversity security, because of this focus on state-centred concepts of security, seem to reinforce this model of reference, but at the same time are affecting issues of gender and environmental security. The global economic, political, ideological, and military lobby is responsible for numerous obstructions in the UN against achieving universal legally binding agreements on the three conventions negotiated at Rio de Janeiro in 1992 (UNFCCC, UNCCD, and CBD).

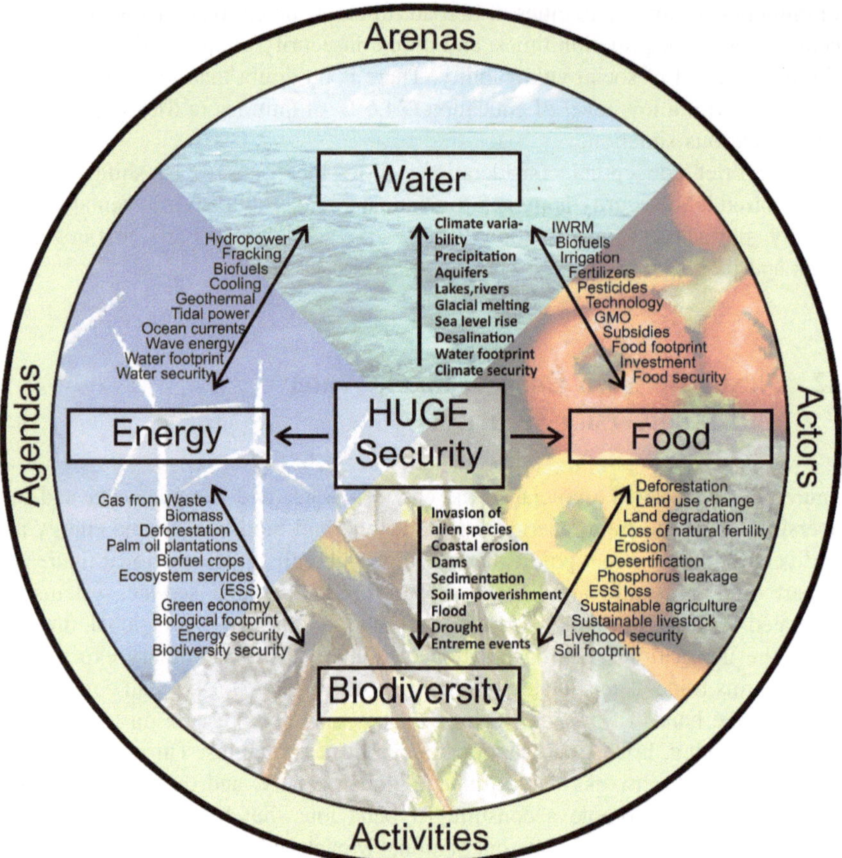

Fig. 6.3 WEF&B security nexus in a neo-liberal arena with actors favouring a *business-as-usual* approach. *Source* The author

6.3 The Case of Mexico

6.3.1 The Political and Socio-economic Context

The WEF&B security nexuses in Mexico are influenced by its specific historical and political context. The 1910 revolution disrupted the economy of Mexico. Ten per cent of its population was killed and its institutions, capital stocks, productive systems and systems of food production were destroyed. The policy of post-revolutionary governments concentrated a growing population in the environmentally fragile environments of the drylands. The first presidents, former generals, came all from the northern arid regions. They invested in extended irrigation systems in their desert and semi-arid regions, and later also in Jalisco, Michoacán and the Bajío in the centre of the country. During the great depression, the last general, Lázaro Cardenas (1934–1940), nationalized the oil industry, distributed land to peasants (*ejidos*), promoted agricultural cooperatives, and created an agricultural bank (*Banrural*) and the Polytechnic Institute. These policies stimulated growth rates in Mexico. During and after World War II, government policy and investment favoured industrialization with import substitution.

The inhabited valley of the capital grew outwards over a former lake, producing water scarcity, and new industrial parks were established in neighbouring states around the capital, which today have about thirty-seven million inhabitants. Although this wider metropolitan area is connected by highways, public buses and a limited train service that facilitates the transportation of people and merchandise within this central region. Simultaneously, two additional urban agglomerations emerged in the north around the state capitals of Guadalajara and Monterrey. The so-called Mexican miracle of industrialization that had lasted four decades ended abruptly in 1982 with a serious debt crisis. The data on its economic performance, measured in terms of *gross domestic product* (GDP), document per capita increases from 1922 to 1981 of 4.85 %; from 1982 to 1985 of only 0.49 %, from 1985 to 2012 of 2.44 %; in 2013 of 1.1 %; 2014 of 2.1 % and 2015 of 2.2 %. This included the economic crisis in 2009 when there was a fall of 9.4 % (Fig. 6.4).

The living conditions of the people had improved over six decades, but since 1982 the prevailing neo-liberal model has concentrated wealth in a few rich families. This new model of accumulation has generated a peak of internal inequity. Mexico's level of inequity is one of the most significant in the world, with a Gini index of 0.472 in 2013 and 0.5 in 2014 (OECD 2014a, b). This means that the 10 % of the Mexican population at the richest level owns 28.5 times as much wealth as the 10 % at the poorest level (Coneval 2013).

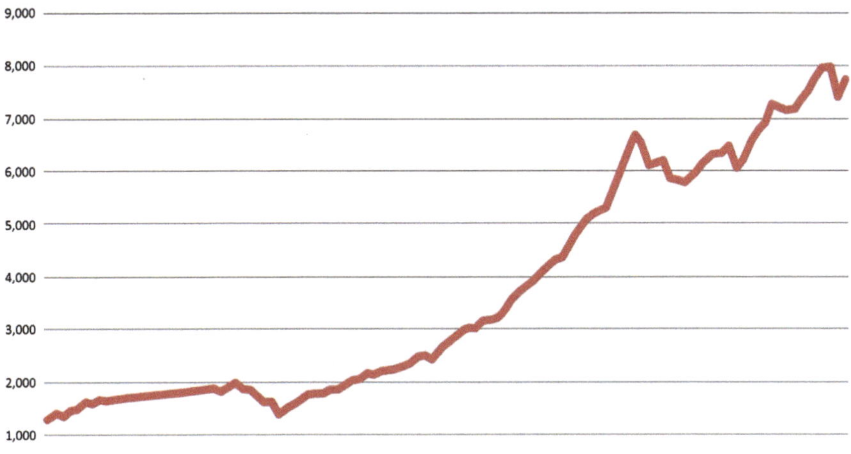

Fig. 6.4 GDP per capita 1900–2010 in Mexico. *Source* PUED–UNAM (2013)

During the phase of fast growth and crisis, a rising population and better living conditions produced a chaotic development process that caused a vicious circle in the environment (Fig. 6.3). Deforestation using slash-and-burn methods destroyed forests, biomass, and biodiversity. It also increased GHG emissions. Fewer forests, and changes in land use from natural to agricultural use, together with extensive grazing by livestock, reduced CO_2 capture and affected the natural fertility of the soil. More chemical fertilizers and pesticides were used to compensate for falling yield, and this polluted soil and water. Agriculture, with its extensive pollution, remains the most important user and polluter of water (Oswald Spring 2011). Only one official standard exists to reduce agricultural and livestock pollution (Pérez 2011), and its legal implementation is weak.

Deforestation of natural areas has also changed the pattern of evapotranspiration, and GHG emissions and climate change have produced more irregular precipitation. Water scarcity and pollution, and 18 % less rain in the drylands (Rosengauss 2007), have caused overexploitation of aquifers, the intrusion of sea water (Fig. 6.5) into coastal aquifers, and failed harvests (Conagua 2014). Scarcity of food, lack of government support to help rain-fed small-scale agriculture, and disasters have pushed peasants and their families into a survival dilemma. Entire families have decided to migrate to cities or abroad (Sánchez Cohen et al. 2011; Oswald Spring et al. 2014). Rural–urban migration has contributed to an already chaotic urbanization with slums, and processes of industrialization have concentrated people in the marginal areas of megacities. The changes from primary to secondary activities and, after the crisis of 1982, increasingly to tertiary or service activities has produced a high level of informal labour conditions and increased urban poverty with no social security. The service sector increased from 52 % in 1970 to 62 % in 2010. The opportunities for study and jobs declined due to the neo-liberal policy of a

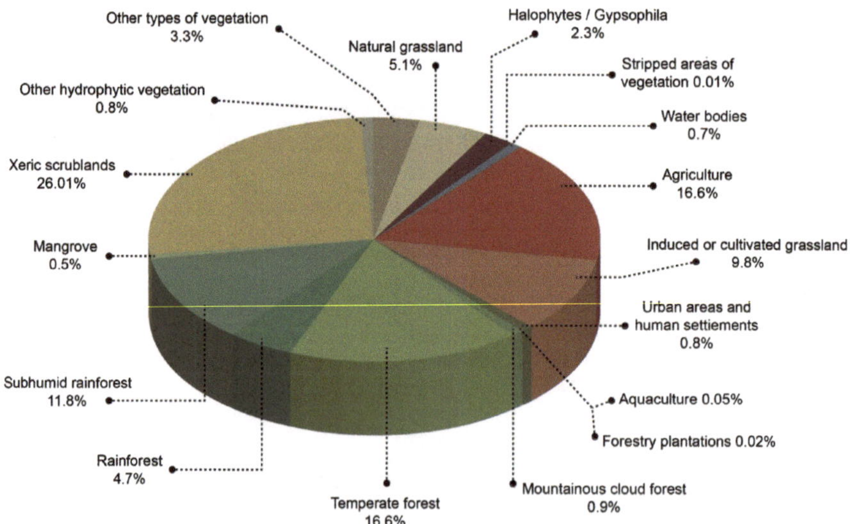

Fig. 6.5 Land use and land cover changes. *Source* CCI (2012: 4)

progressive privatization of education and public services. In 2013 many of the eight million *ninis* (no school, no job opportunities) were involved in organized crime.

During the years of crises people were primarily occupied with their own survival, and in 2000, after seventy-eight years of single-party rule by the PRI (*Party of Institutional Revolution*), the citizens voted for the conservative opposition party PAN (*Party of National Action*). When in 2007 the tortilla price skyrocketed, and ideological control by a duopoly of television companies was failing, people took their protest to the streets. The response of the government was—within the framework of the Merida Agreement with the US—to declare a war on drugs, and military expenditure tripled on the cost of social expenditure.[7] Criminal activities diversified (extortion, kidnapping, rape, robbery, trafficking, etc.). The US Department of Homeland Security (2014) estimates that these criminal activities amounted to about US$400 billion in America alone. Most of this money is laundered outside Mexico, in the fiscal haven of the Cayman Islands, in stock markets, by transnational banks, in real estate development, tourist facilities, commercial malls, etc. in the USA. The economic stability of Mexico and of other Latin American countries is partially linked to these criminal activities and money laundering.

This brief assessment of the linkages between natural conditions and social and political developments in Mexico demonstrates the complexity of the nexuses between WEF&B security. It points up the increasing social vulnerability and the hidden interests behind discussions about nexuses related to national security. In Mexico, from 2006 to 2012 the drug war caused military defence spending to increase at the cost of social and educational expenditure. Most armaments were purchased from the US, but also from Belgium, Germany, the Czech Republic, Spain and Italy. From 2006 to 2012, the drug war in Mexico is estimated to have caused 83,000–150,000 deaths; 27,000–28,000 persons disappeared, and between 250,000 and two million persons were displaced (Piñeyro 2012). According to the Informe Bourbaki (2011), 80 % of the people killed were murdered by organized crime, mostly young men of whom 21 % belonged to military forces and government authorities, and 56 % are unknown and considered as "socially disappeared". Among the official dead, 43 % are classified as being the result of social repression and of the elimination of social leaders, 31 % as drug offences, and 26 % as other cases. The Bourbaki Report concludes that it is not a war against drugs but a civil war, and Jaso (2013) referred to violations of human security in terms of *freedom from want*, but also of *freedom from fear*. The drug war has also increased corruption. The army and the navy carried out police activities and several officials were bribed by organized crime, even to let the most important drug lord escape from a high security jail.[8] Corruption by the drug rings affected all sectors of the

[7]This has not changed with the present government under Peña Nieto, as can be seen in the federal budget for 2015. This proposes increases in spending of 6.4 % for the navy and 5.8 % for the army, but only 0.8 % for education and 0.1 % for health, and a reduction of 20.5 % for agriculture (Budget approved for 2015).

[8]The expenditure on military jails tripled between 2006 and 2012 because of corruption among the military involved in the drug war, bribed by organized crime and fined by military tribunals.

economy and money is also laundered by the business sector. In 2014 Transparency International placed Mexico in 103rd place out of 175 countries. The Mexican government and the army paid additional costs for this loss of transparency, but globally the legitimization of a high level of military spending, previously justified by the cold war, has now been replaced by the US-defined war on drugs and the war on terror.

The war on drugs has therefore increased military expenditure to the detriment of human security. But this militarization has affected not only human security, but also environmental security. In Mexico water, energy, food and biodiversity were securitized because they were seen as threats to national security, as in the example of illegal timber logging, which is carried out by organized crime.

6.3.2 *The Nexus Between Water, Energy, Food and Biodiversity Security in Mexico*

Mexico has 1,964,375 km^2 of land area: 1,959,248 km^2 are continental and 5,127 km^2 are islands. The topographical conditions and differences in altitude result in different ecosystems, of forests, jungle and wetlands. Important changes in land use from forest to agriculture and greater climate variability affect ecosystems, and produce direct pressure on water and biota (Fig. 6.5). Mexico has an unbalanced supply of water when it comes to time, regions and social classes. Between 72 and 78 % of precipitation occurs during the monsoon season (June–September). Northern and central drylands receive less than one-third of the precipitation, contain 77 % of the population, and produce 79 % of the GDP. This imbalance between water supply and consumption requires a transfer of water from neighbouring basins, and this has triggered numerous conflicts and environmental destruction, and a deterioration in water quality and human and environmental security. The transformation of natural ecosystems into urban ones has aggravated environmental and social risks. The nexuses of WEF&B exercise pressure on each resource because of scarcity, exhaustion and pollution. However, harmful feedback within each system and between the natural and human systems has reinforced the negative impacts, with complex societal outcomes. The nexuses between the different resources are multifaceted.

6.3.2.1 Water–Energy Nexus

Energy demand rose in Mexico between 1994 and 2004 by 4.7 % per year and it is estimated that it will increase by 4.8 % by 2015. About 26% of electricity is produced with renewables (Sener 2012). Table 6.1 indicates the existing status and the potential of renewables; hydro, wind and geothermic energies are most important for Mexico. The estimates for concentrated solar energy are still low, but

Table 6.1 Installed and projected renewables in Mexico

Renewable	Installed (2012) in GW	Potential in 2024 (GW)
Wind	283	40,000
Water	990 (11,272 Conagua)	53,000
Geothermal	11.7	500–1,000
Photovoltaic (solar)	100	312
Solar concentrator	2.5	30
Biomass	83	134
Ocean	0.5	2

Source Sener (2012)

given the high solar potential in the extensive drylands Mexico has a high solar energy potential overall. Biomass energy from forests, waste, animal manure and urban organic waste may generate important sub-products for soil restoration and organic agriculture, and so relates to the nexus between energy and biodiversity. The national oil company Pemex estimates it will extract 2,937 MBD of oil and 8,061 MCFD of gas between 2012 and 2026, with an investment of 339.9 billion pesos (US$ 28.325 billion): 37.9 % for the exploitation of existing oilfields; 22.7 % for exploration of new fields; 17.7 % for deep-sea projects, and 7.6 % for two fields of shale gas (Sener 2012: 20). All these activities mean costs, contamination, GHG emissions, and high risks in the Gulf of Mexico and in the shale gas projects, and will have a negative impact on biodiversity and the public image of renewable energy in the country. Between 1990 and 2010, GHG increased by 34 % and those from government energy monopolies by 58%. Waste, managed by local government, private concessionaires, and also industry increased them further. GHG emissions related to changes in land use (deforestation) were stabilized, because most of the available forests have been destroyed. Nevertheless, deficiencies in the rule of law and the shortcomings of the judicial system have promoted illegality, impunity, and loss of biodiversity.

Among the renewables, hydropower is the most important in Mexico and dams were constructed during the period 1970–1980.[9] The displacement of poor indigenous inhabitants and peasants and the lack of fulfilment of government commitments have increased the opposition to new dams. There has also been lack of transparency on the costs of these important public works, long-term amortization, and the restriction of international funds. Alternatives for producing sustainable electricity are small dams and hydroelectric turbines in river beds.

[9]The most important dams are Chicoasén, La Angostura and Malpaso in Chiapas and Infernillo in Guerrero, all in indigenous and very poor regions.

Agriculture has reduced its GHG emissions by almost half, as intensive livestock-rearers manage the manure with the support of programmes funded by the *clean development mechanism* (CDM). Finally, the upcoming technology of fracking and shale gas in the deserts of Chihuahua, Tamaulipas and Nuevo León uses up scarce water resources. There is a danger that these fossil water reserves may be destroyed through the use of toxic chemicals. It is also a region with intensive conflicts between drug cartels on the trade routes to the North American drug market. The increase in GHG was complemented by rhetorical references to Mexico as a renewable country. Future policy is still centred on fossil fuels at the cost of human security, especially *freedom from hazard impacts*.

6.3.2.2 Water–Biodiversity Nexus

"In Mexico between 3.5 and 5 million hectares of temperate and tropical forests have been lost during the last decade. The estimated annual deforestation rates in Mexico range from 0.5 to 1.14 % since the early 1990s to 2000" (Commission for Environmental Cooperation n.d.: 2). Cloud forest, mangroves, sub-humid rainforest and rainforests are the ecosystems that have been most widely destroyed, while xeric scrubland is expanding due to drier conditions. The reduction in natural vegetation has a large impact on evapotranspiration, heat flux, the albedo and the temperature, and reduces the absorption of CO_2 (Huete et al. 2002). Besides deforestation, the biodiversity of Mexico is threatened by changes in land use and by human settlement. Human activities have drastically modified the original vegetation cover and landscape of Mexico (Fig. 6.5), affecting water infiltration, aquifer recharge, and *ecosystem services* (ESS). The drier conditions in urban settlements cause the quality of air to deteriorate, as well as the run-off of water and human health, and they reinforce the negative outcomes of climate change.

Healthy soils are crucial for biodiversity, yet 88 % of the land is affected by water erosion (CCI 2012). Because of increasing temperature and longer droughts in the drylands, the moisture in the soil is declining and food production and natural ecosystems are under stress. It is estimated that climate change will reduce rainfall on the land by 12 % by 2100 (CCI 2012). Cook et al. (2010) point to the key importance of evaporation, which will "increase the percentage of global land area projected to experience at least moderate drying by the end of the 21st century from 12 to 30 %". Romm (2011) forecasts that "precipitation patterns are expected to shift, expanding the dry subtropics. What precipitation there will probably come in extreme deluges, resulting in runoff rather than drought alleviation. Warming causes greater evaporation and, once the ground is dry, the sun's energy goes into baking the soil, leading to a further increase in air temperature".

Rising sea level and a high level of groundwater withdrawal from aquifers by dryland agriculture, growing cities, and tourism in the coastal areas are causing the

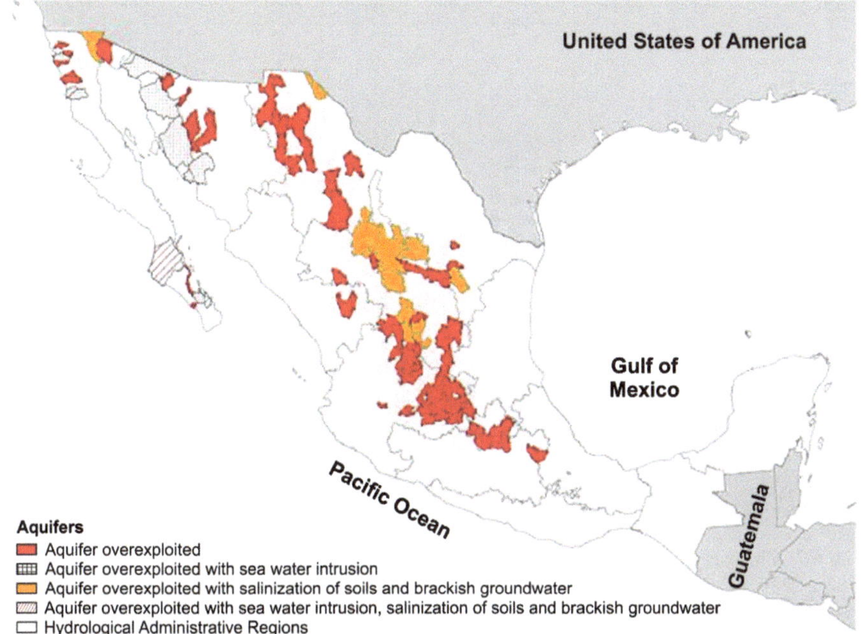

Fig. 6.6 Intrusion of seawater into the aquifers and brackish water on the soil. *Source* Conagua (2014)

intrusion of seawater into the aquifers (Rangel et al. 2011). In the north and north-east, intensive irrigation with brackish water has further affected the quality of the soil (Fig. 6.6). Additionally, water pollution is increasing water scarcity. Finally, water quality has deteriorated due to an ageing tap water infrastructure and natural pollution by arsenic (Pacheco et al. 2011) and fluoride. Around the metropolitan areas of Mexico City, Guadalajara, Monterrey, the Northern Gulf and the Central Gulf, the pollution is aggravated by industrial activities and a concentration of people,[10] and by a lack of or inefficient management of treatment plants.

[10]The total suspended solids, the biochemical demand for oxygen and the demand for chemical oxygen affect the coastal zones from Colima to Guerrero on the Pacific, the south of Veracruz and Tabasco on the Atlantic, and the rivers of Santiago-Lerma, Bravo and Soto La Marina (Arreguín et al. 2011). The quality of water in most cities and rural areas is not safe and most inhabitants buy water jugs, not always of the expected quality. INEGI (2010) reports drainage coverage of 86 %, but gastro-intestinal illnesses, viral diseases, poisoning, typhoid and paratyphoid are still common diseases in Mexico. The distribution of oral rehydration, immunization, clean water and better hygiene in homes and school programmes has reduced mortality from acute diarrheal diseases, especially in urban areas. But there are still important differences between states, and between urban and rural areas: Tabasco has a mortality rate per 100,000 inhabitants of 0.93, while in rural areas of Chiapas the rate is 18.03 (Cortes/Martin 2012).

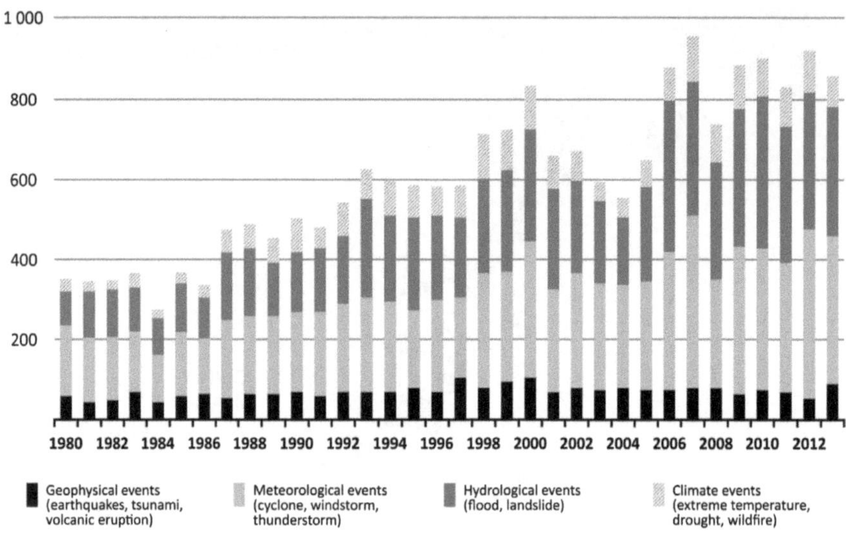

Fig. 6.7 Changes in natural hazards in Mexico. *Source* CCI (2012: 14)

Climate change has contributed to increased and more severe extreme events and Mexico is highly exposed to cyclones from the oceans on both sides (Fig. 6.1). Mexico is equally exposed to other types of hazard (Fig. 6.7); since 1999 floods and droughts have intensified, together with wildfires and heatwaves. These mostly climate-induced hazards are increasing dual vulnerability. To improve its human security, both the government and the people must improve their capacity for adaptation, since people will face progressively more difficult conditions.

Most people live in the drylands and the major productive activities occur there. This is where climate change will affect more seriously, and affect humankind and the economy. Agriculture in Baja California Norte is under serious stress as the seepage from the Rio Colorado disappeared when the US lined the All-American Canal. Water scarcity is also severe in the megacities of the drylands, especially in the *Metropolitan Valley of Mexico City* (MVMC). This has overused its six aquifers, considered among the most exploited on earth (Oswald Spring 2011), causing negative impacts from subsidence of up to 50 cm per year. This is affecting infrastructure, drainage and the water supply system, roads and houses. The MVMC imports one-fifth of its water supply from neighbouring states, and this creates water conflicts with the indigenous Mazahua people. The chaotic growth of the megacity has also destroyed the former lacustrine ecosystem, and during the monsoon, the MVMC is permanently exposed to floods while during the dry period there is a severe scarcity of tap water.

Nevertheless, these natural conditions do not affect all citizens in the same way, and social and economic differences, water reservoirs in houses, and money to buy water from tanks is a privilege of the wealthy. Finally, the poorest people who have recently immigrated to the MVMC live mostly in high-risk areas (ravines, unstable soils, steep slopes exposed to landslides, flood-prone zones), because the price of land set for real estate for the bourgeoisie is beyond the reach of newcomers. Again, environmental security is worsening social vulnerability and vice versa, and the most socially vulnerable are exposed to hazards, limiting their human security in terms of freedom from fear and freedom from hazard impacts.

6.3.2.3 The Water–Food Nexus

Development policy in Mexico has produced a great imbalance between water availability, population settlements and productive activities, further aggravated by time restrictions on the availability of water during the dry season. Seventy-seven per cent of Mexico's water is used in agriculture, 13 % in the domestic and service sector and 10 % in industry. The low efficiency of irrigation in the arid north as used by export-oriented agribusiness creates a sectoral imbalance, and the lack of treatment and low rate of recycling reduces and pollutes the existing water resources. Most rivers are converted into drainage for sewage and waste, where leachate pollutes groundwater and soils and limits the development of plants and animals.

Agriculture is thus a key user and polluter of water in Mexico, and irrigation efficiency reaches only 48 % (Palacios/Mejía 2011), since most irrigation relies on open channels. In 2010, the lack of water and an ongoing drought forced farmers to reduce their irrigated areas by 6 % annually. Nevertheless, half of the water in agriculture could still be saved through further technological improvements (satellite-managed irrigation, closing of channels by pipes, the levelling of agricultural fields, changing water-intensive crops, greenhouse production, and renovating old irrigation systems). However, a lack of credit and investment because of the unstable prices of cash crops, and a system of inadequate subsidies for diesel, water and electricity in water-stressed regions has held back improvements in irrigation.

Maize is the basic food crop and Mexico is its region of origin and adaptation. Mexico produces around 23 million tonnes of maize, of which 57 % comes from rain-fed fields. Climate change projections indicate that between 13 and 27 % of this area of maize may be lost (CCI 2012). This will especially affect poor peasants who rely on rain-fed agriculture, and they often migrate to cities or abroad in order to survive.[11] Since 1982, the government has also drastically reduced support for

[11]Today 2.7 million productive units (66 %) belong to peasants cultivating less than five hectares. Despite the negative climate conditions, the yield doubled between 1990 and 2007, reaching an average of 2.82 tonnes per hectare (Robles Berlanga 2010). This is the result of the so-called *cero labranza*, meaning that peasants produce their crops without chemical fertilizers and use animal manure and organic waste to improve soil quality. This traditional way of producing maize, with seeds carefully selected from the previous harvest, explains the success and the variety of the germ plasm of maize, as well as its resistance to adverse climate conditions and the maintenance of soil fertility in mountainous areas. During the NAFTA process, Mexico negotiated a protection clause for the importation of maize, but never implemented it nor charged import taxes, and so has lost taxes worth more than US$27 billion since 1994 (SHCP 2011); this has seriously affected small-scale producers.

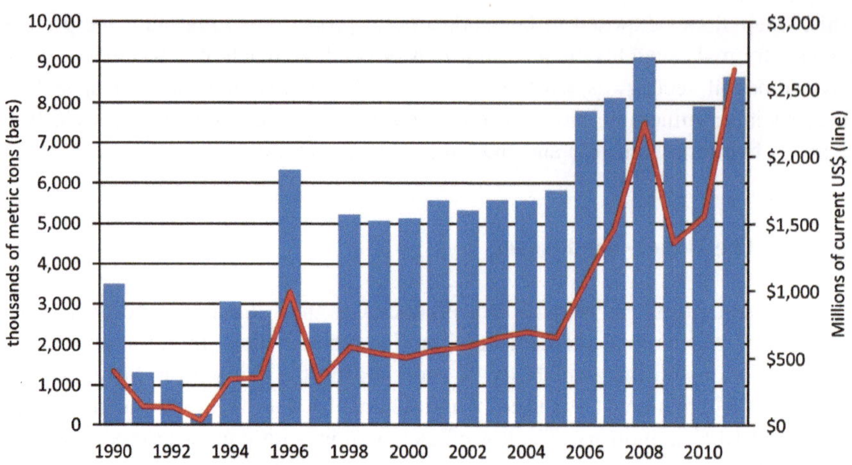

Fig. 6.8 Maize imports to Mexico. *Source* Turrent et al. (2013: 5)

small-scale farmers and oriented subsidies towards export-oriented agribusiness, under pressure from powerful lobby groups. The second factor in the destruction of the peasant sector is the *North American Free Trade Agreement* (NAFTA). The Mexican government changed its constitution and permitted the renting, selling and association with private business of *ejido*[12] land, so that the peasants lost the protection of their land rights. Since the 1990s, the government has not compensated its peasants for the import of heavily-subsidized maize from the US, which has destroyed internal maize prices and local market structures and affected local maize production (Bartra 2012). Mexico is importing increasing amounts of maize (Fig. 6.8), mostly from the US (USDA 2013), and over 90 % of these imports are from *genetically modified organisms* (GMOs), creating an additional threat to biodiversity security.[13]

Wise (2012: 169) calculated the losses in the income of maize farmers due to this neo-liberal policy: "…the dumping-level price was more than $6 billion over the nine-year period, or $730 million per year (in constant US dollars of the year 2000). Losses exceeded $11 billion since 1990, with the highest losses in 1993, and in 1999 and 2000 when dumping margins exceeded 30 %. From 1997 to 2005 producers lost an estimated $38 per metric ton of corn, or $99/ha per year. For most years, per hectare losses were between $50 and $100. In 1993, 1999, and 2000, losses exceeded $175/ha". This unfair commercial behaviour and with the

[12]*Ejido* is the land peasants struggled for during the Mexican Revolution in 1910 and later was redistributed -sometimes collectively- from the government through land reforms.

[13]Mexico imported in 2015 about 25.3 % of its maize consumption, 59.0 % of wheat and 88.4 % of soya beans (INEGI 2015).

simultaneous reduction of government subsidies in Mexico increased the precariousness of its peasants.

The lack of a government vision of a future for the country and for the most vulnerable was further threatened by the import of genetically modified corn, which created additional risks for the biodiversity of maize (Oswald Spring 2009b). Because of corruption and the wish to provide benefits for grain importers (enterprise called Gruma), Mexico lost its national food security and also a part of its food sovereignty (Turrent et al. 2013). Several Mexican governments also succumbed to North American pressure and to the structural adjustment policy of the IMF. The links between natural, socio-economic and political conditions, but especially the reduction in wages due to economic crises since 1972 and the changes in traditional food patterns, have also altered the traditional diet, causing serious health impacts (diabetes, cardiovascular diseases).[14]

6.3.2.4 The Energy–Food Nexus

The energy–food nexus relates to the production of biodiesel and ethanol from agricultural crops. The US provided large subsidies to boost the production of ethanol from production of maize. Figure 6.9 indicates the increase in the production of maize destined for ethanol production in the US. This produced a price rise in 2007, but the outcome in terms of GHG is not favourable.[15] The energy–food nexus has also altered food prices. Maize prices were stable for decades, but between 2007 and 2012 the growth in ethanol production from maize increased its price on the Chicago stock exchange from US$131.27 to US$317.15 per tonne.

Qiu et al. (2012) support the thesis that the fundamental market forces of supply and demand are the main drivers of food price volatility, where increased biofuel production is causing short-term price rises, but is not producing a price shift in the long run. Alternatively, Merkusheva/Rapsomanikis (2014: 4) affirm that "the total demand for maize, that is the aggregate demand by ethanol and food consumers, is also linked, corresponding to the demand for ethanol with segments that are less elastic determined by the price elasticity of the demand for ethanol, and that for food and feed". The same authors maintain that automotive engine technology and US biofuel policies are establishing a non-linear relationship between oil and

[14]Ensanut (2012) indicates that Mexicans consume the double of the recommended intake of sugar, but their diet is deficient in cereals, legumes, fruits and vegetables. Only 14 % of small children are adequately fed and the abuse of soft drinks and the lack of exercise is increasing obesity.

[15]This biofuel policy has produced additional threats related to climate change: the 2012 drought affected 75 % of the maize production in Iowa, Illinois, Kansas and Nebraska, and in the US as a whole with an average of 56 %, and corn production declined from 379 to 274 million tonnes, that is, 100 million tonnes less than expected. Concerns over food security forced China to abandon its ethanol programme. Mexico, with a deficit of maize, never seriously started to produce bioethanol, but the government learnt from the price rises that food security prevents conflicts and protests.

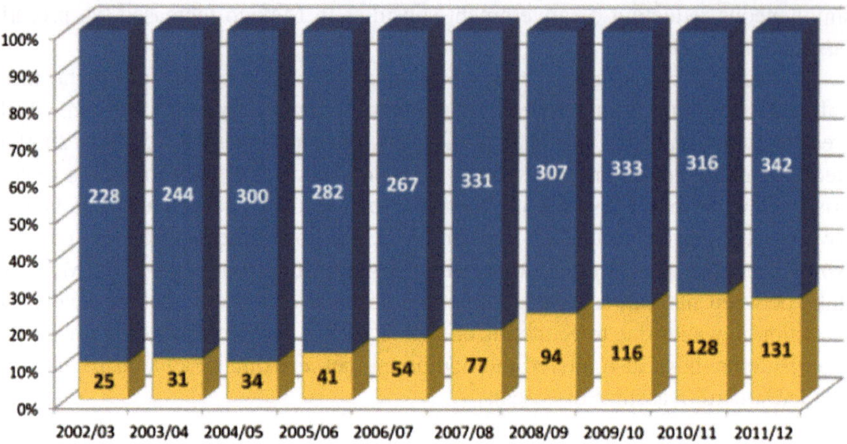

Fig. 6.9 The proportion of maize used for ethanol production (in *yellow*) in the US (2002/2003 to 2011/2012). *Source* Ministry of the Economy (2012: 9)

ethanol processes. Serra et al. (2011) confirmed this non-linear relationship with prices, because energy prices affect the price of maize through the ethanol market. The empirical evidence indicates that the demand for maize from the subsidized ethanol industry not only affects this crop, but because of substitution also affects other food products, especially rice.

6.3.2.5 The Food–Biodiversity Nexus

Mexico has ten distinct ecosystems with 958 endemic species of fauna and 5,161 of flora (Conabio–UAEM 2006: 155). This biodiversity is threatened by population growth, growing demand for food and water, deforestation, and human settlement. The food–biodiversity nexus is directly linked to changes in agricultural land use, soil management, agrochemicals, genetically modified seeds, and livestock production. Half of the land area is used for agriculture and livestock, and tropical jungle and forests are being destroyed without providing food security for Mexico. The result is an unsustainable level of soil management, pollution, and intensive use of water. These environmental and social factors have resulted in a high level of GHG emissions and have created more than 400,000 conflicts over land and water.[16]

[16]One example of how the lack of water for people, productive activities and agriculture was dealt with is provided by the Hermosillo Valley, the capital of Sonora. The three levels of government constructed an aqueduct from the Novillo dam with no negotiation or any environmental impact study. The traditional water rights of the indigenous Yaqui people were overruled. This arbitrary behaviour by the government has created a critical ongoing water conflict in Mexico.

Biodiversity is also threatened by a growing population's demand for food,[17] but Mexico produces a great variety of different food products.[18] The food–biodiversity nexus is most seriously affected by drought. "Since the second half of 2010 a significant lack of rain in 19 states of Mexico became a severe drought causing losses over 15,000 million pesos" (CCI 2012:15). This amount represents 6.39 % of GDP in the agricultural and livestock sector. Harvests of corn, beans, and vegetables severely declined and livestock perished. About 2,350 communities and almost two million people were affected. From 2003 to 2013, the drought between 2011 and 2013 resulted in an extreme situation, with severe droughts in more than 70 % of the national territory. In 2011, drought (Figs. 6.10 and 6.11) caused the loss of 1.8 million hectares, almost 5 % of the arable land (CCI 2012), and this increased dual vulnerability. All these developments led to poverty for many small farmers; this was also partly due to the loss or depletion of important natural resources.

6.3.2.6 The Energy–Biodiversity Nexus

The last nexus is the link between energy and biodiversity. In Mexico this relationship is mostly indirect and relates to oil and gas exploitation. When the oil boom began in Tabasco in the late 1970s, the drilling of wells, the construction of new roads, and a World Bank policy of extensive livestock-rearing in the humid tropic zone caused the destruction within a decade of 92 % of the tropical rainforest. It also caused the wetlands in this region to dry out (Barkin/Zavala 1978). In addition, the processing of crude oil in petrochemical plants has affected air, soil, water, aquaculture, and fish stocks in rivers, wetlands, and deltas. Acid rain has affected plants, animals and infrastructure. Most electricity in Mexico is still produced from low-quality fossil fuel with a high carbon footprint. The two

[17]Population has increased from 20 million in 1940 to 120 million in 2013. In 2011, however, the fecundity index fell to 2.17, and the number of children per woman decreased on average from 6.78 in 1960 to 2.28 in 2010. There is greater population growth in rural areas, while urban growth can be explained by rural–urban migration trends; since 1960, the urban population has exceeded the rural. Economic activities and wealth are concentrated in the industrial and service sectors in urban settlements. In 2012 agriculture produced 6.7 % of GDP and employed 13.5 % of the workforce; in the previous year, poor climate conditions had reduced primary GDP by 2.6 %. These data illustrate a sectoral imbalance, with low salaries and a high dependency on climate factors.

[18]Food items: grains and seeds (white, red, blue, and yellow maize, cacao, amaranth, chia, peanut, sunflower, pine); pods (beans, gourds, mesquite); vegetables (squash, quintoniles); leaves (purslane, chaya); fruits (green and red tomatoes, varieties of potatoes, chayote, chilacayote, mamey, avocado, custard apple, papaya, soursop, sapodilla, plums, lechuguilla, chia, guava, dragon fruit, tuna, nopal, chico, tejocotes); roots (sweet potatoes, yam beans); flowers (hibiscus, squash flower, yucca); hot peppers (guajillo, habanero, jalapeno, chipotle, chile de arbol, etc.); herbs (coriander, epazote, vanilla, annatto dye, onion, sacret leaf [*Piper Auritum*], pumpkin, chipilin); mushrooms (huitlacoche); insects (chinicuiles, grasshoppers, escamoles-ant eggs, jumiles); meat (turkey, deer, xoloitzcuintle, birds, fish, shellfish, shrimps, sea fruits and many other animals), etc.

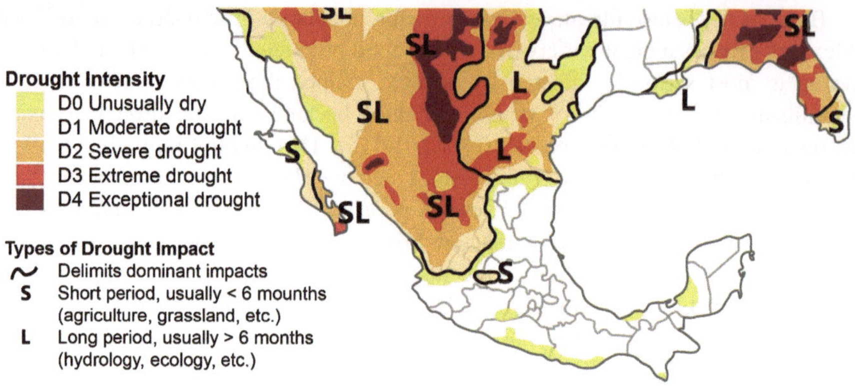

Fig. 6.10 Drought monitoring, October 2011. *Source* CCI (2012: 179)

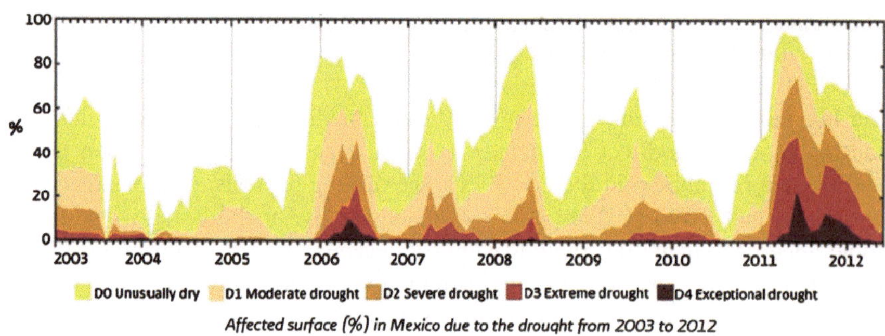

Fig. 6.11 Land area (by percentage) affected by drought 2003–2011. *Source* CCI (2012: 17)

state-owned companies, the oil company Pemex and the electricity company CFE, are both primary polluters of air, soil and water, and have a negative impact on agriculture, flora and fauna.

Figure 6.3 above indicated numerous negative linkages between environmental and human factors, and these both contribute to dual vulnerability and increase human insecurity by attacking *freedom from want* and *from hazard impacts*. In terms of want, the concentration of wealth in a small oligarchy was only possible because of the loss of workers' purchasing power. This was caused by a deterioration in salaries, the privatization of public services and the structural adjustment policy of the IMF following the imposition of the neo-liberal model. The Gini index of 0.50 (European Union 0.29) indicates a very high level of internal inequality in Mexico. The evolution of wealth distribution shown in Table 6.2 indicates that wealth was further concentrated during the crisis years, when 10 % of the richest began to own one-third of the national wealth, while half of them owned 28.4 % (OECD 2011, 2014a). As a result, 60 % of Mexicans are forced to work in the informal sector with no regular income (OIT 2012), and live in poverty.

Table 6.2 Concentration of wealth during the recent crisis in Mexico (by percentage)

Decile/year	2000	2002	2004	2006	2008	2010	2012
I–IV	25.3	27.0	26.9	26.7	27.6	26.7	28.4
VII–IX	36.1	37.4	36.9	36.8	36.7	37.0	37.7
X (richest 10 %)	28.6	35.6	36.2	36.5	35.7	36.3	33.9
Total	100	100	100	100	100	100	100

Source INEGI (2012)

This social vulnerability is further aggravated by environmental threats. Poor people in the south and south-east are heavily affected by hurricanes and in the north by serious droughts. Most of the vulnerable people live in precarious shelters and can lose their belongings when exposed to extreme events. Dual vulnerability thus increases the threats to people who suffer from a lack of education, income, and social security, as well as access to resilience measures and disaster risk reduction and management. New extreme events often turn into disasters causing loss of life, loss of livelihood and loss of productive opportunities for these vulnerable people.

This situation is linked to internal factors such as government inefficiency, the lack of a trained and professional bureaucracy, the low level of school education, a limited level of innovation, a high level of corruption, the fact that the drug war triggers a high level of public insecurity, capital flight by the oligarchy, corruption, and the absence of the rule of law (Bailey 2014). In addition, since 1994 external pressures such as unfavourable terms of trade, assembly industry called *maquila*, and unequal relations within NAFTA have limited the creation of well-paid jobs and social well-being though stable employment. Not only that, but tax evasion and corrupt processes of privatization have deprived the Mexican state of much-needed financial resources to support the poor and to modify the unequal income structure. These structural conditions of poverty, corruption and inefficiency have reduced Mexican citizens' trust in their government to only 37 % (Latinobarómetro 2013: 19).

6.4 Complex Emergency

Complex emergency is characterized by the International Red Cross as a displacement of people due to disaster, extensive violence, and loss of life, with widespread damage to society and the economy. Conditions of Mexico belong in this category: frequent extreme events, economic crises, and organized crime, all resulting in dual vulnerability, displacement, and death. Figure 6.12 shows the drug, human and other illegal trafficking routes through Mexico that supply the US with undocumented workers and illegal products and services. The amount of

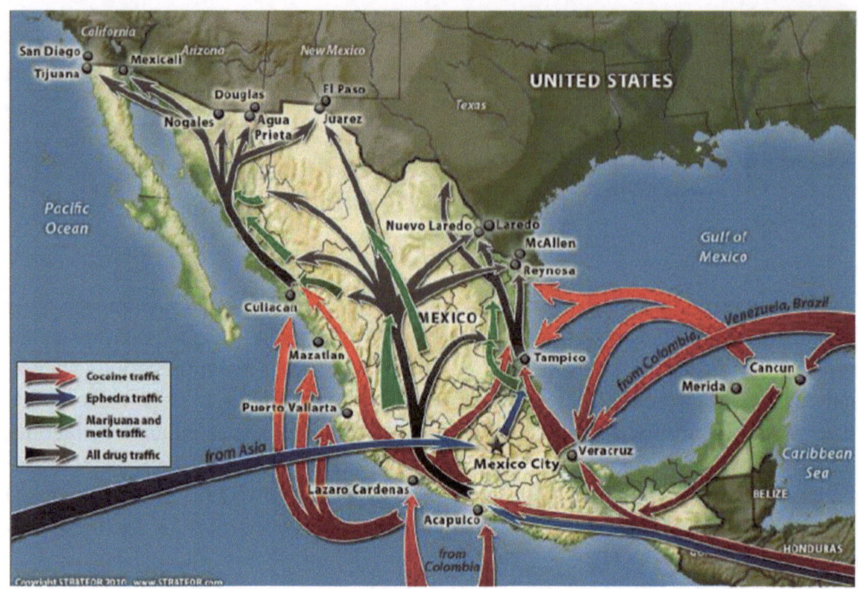

Fig. 6.12 Drug trafficking though Mexico. *Source* Informe Bourbaki (2011: 15), based on Stratfor (2010)

money involved in this traffic benefits the global oligarchy and the leaders of the Mexican drug cartels, but has also contributed to the stabilization of the economic crises (Murphey 2013).

Bailey (2014) maintains that Mexico is suffering a triple crisis, which has created a high level of insecurity. There is no social contract and the numerous agreements signed by the government, political parties and business have always adversely affected workers and the informal sector but benefited the oligarchy. The second structural problem is the lack of reform of the judicial system and the police. Today most local authorities and police are infiltrated by organized crime. The third problem relates to political parties and the electoral system. A fourth problem is corruption within the government, enterprises, and the corrupting capacity of organized crime. If the economic benefits of criminal activities are not addressed, there will be no chance of controlling illegal activities. While many leaders of the cartels have been put in jail or killed, they have always been replaced by someone else, or they escape from prison. There is a fifth, more structural, problem linked to the neo-liberal model of development. This has reduced Mexico's growth rate and concentrated wealth in a small national and international oligarchy through dubious processes of privatization, free trade agreements, and structural adjustment programs imposed by the IMF. The Mexican bourgeoisie has often initiated the flight of capital and aggravated the economic crises.

6.5 Conclusions

With respect to the research questions, this chapter has first described dual vulnerability and the complex linkages between environmental and social aspects. With respect to the initial hypothesis, the analysis of the nexuses between WFE&B launched by WEF indicates, in the case of Mexico, a hidden oligopolistic expansion of national and global capital at the cost of the majority of the people. This oligarchy has increased their wealth alongside with legal business dealings through forced privatization processes, money laundering, capital flight, tax evasion, bank rescues, and corruption. High-level government employees have often been involved. The hidden security agenda also explains how the members of this oligarchy ally themselves with national governments to reinforce their power structure directly through the military, but also through ideological means, such as the involvement of the most important Mexican television channels in the electoral campaigns of top politicians.

The study shows that a critical analysis of the nexuses between WEF&B security reflects the present serious situation where there are global problems related to the climate system, water, biodiversity, and soil deterioration. This analysis has also shown the usefulness of a human security approach focusing on dual vulnerability for the analysis of the nexuses between water, energy, food, and biodiversity security. This approach may also contribute to reframing policies for dealing with the linkages between these nexuses and overcoming the hidden military and political implications of security. Approaches could be:

1. Mexico displays a high level of biodiversity in maize, and compensation for the peasants who with increasing effort reproduce the germ plasm would conserve this invaluable natural patrimony of humanity. A clear level of government support could protect the germ plasm of this crucial food item. Recovering environmentally fragile regions and investing in food security and sovereignty would reduce dual vulnerability, improve livelihood and nutrition among marginal groups, recover foreign exchange, and create greater social justice for the poorest but most biodiverse regions of Mexico.
2. Indirect ways to reduce the negative impact of food production on biodiversity include a reduction in loss, recycling of waste, reduction of meat intake,[19] improvement of the local food market, and a reduction in subsidies for agribusiness and trade. This would promote sustainable and biodiverse agriculture with a low carbon impact.
3. Faced with drought, the government and agribusiness in the north could promote water-saving irrigation projects. Today, farmers use the saved water reserves to expand their irrigated areas, when the government's intention was to

[19]Seventy per cent of direct GHG in agriculture comes from livestock (Dickie et al. 2014), especially from grazing animals (cows, sheep etc.), and less meat production would reduce these GHG emissions.

transfer these water reserves to domestic needs and grant water security to the people (Sánchez Cohen et al. 2011).

4. Mexico has laid stress on human security in the past at the United Nations (Friends of Human Security). By focusing on basic human needs, the hidden agenda of hard military security could be overcome and the country helped to refocus its development priorities. By centring its public and private investments on development challenges, dual vulnerability could be more efficiently addressed, social inequality reduced, education improved, the science and technology base could be strengthened, and the Millennium Development Goals could be achieved for all Mexicans.
5. From a human security perspective, a securitization of dual vulnerability is proposed as a political tool to reduce environmental and social vulnerability in Mexico, by making these vulnerabilities an issue of utmost importance that requires extraordinary measures by the government and the people affected.
6. At the global level, there is *only one planet Earth* in our solar system where the conditions for life and joy exist. Not only the poor but also the global oligarchy have no alternative but to live here. A shift that enhances human and environmental security by overcoming the dual vulnerability of the majority of the people—in synthesis a HUGE security—may open ways to change the present *business-as-usual* policy towards a transition to a sustainable development (Grin et al. 2010) for nature and humankind.

References

Annan, Kofi, 2005: *In Larger Freedom: Development, Security and Human Rights: The Millennium Report* (New York: UNO).

Arreguín, Felipe; López Pérez, Mario; Marengo Mogollón, Humberto, 2011: "Mexico's Water Resources for the 21st Century", in: Oswald Spring, Úrsula (Ed.): *Water Resources in Mexico. Scarcity, Degradation, Stress, Conflicts, Management, and Policy* (Berlin–Heidelberg: Springer-Verlag): 21–38.

Barkin, David; Zavala, Adriana, 1978: *Desarrollo Regional y Reorganización Campesina: La Chontalpa Como Reflejo Del Problema Agropecuario Mexicano* (Mexico, D.F.: Ed. Nueva Imagen).

Bailey, John, 2014: *The Politics of Crime in Mexico: Democratic Governance in a Security Trap* (Washington. D.C.: Georgetown University Press).

Bartra, Armando, 2012: *Los nuevos herederos de Zapata. Camepsinos en movimientos 1920–2012* (México, D.F.: CNPA, Circo Maya, México, D.F.—Cámara de Diputados—Circo Maya).

Beck, Ulrich, 2009: *World at Risk* (Cambridge: Polity Press).

Beck, Ulrich, 2011: "Living in and Coping with World Risk Society", in: Brauch, Hans Günter; Oswald Spring, Úrsula; Mesjasz, Czeslaw; Grin, John; Kameri-Mbote, Patricia; Chourou, Béchir; Dunay, Pal; Birkmann, Jörn, 2010: *Coping with Global Environmental Change, Disasters and Security—Threats, Challenges, Vulnerabilities and Risks* (Berlin–Heidelberg–New York: Springer-Verlag): 11–16.

Bogardi, Janos; Brauch, Hans Günter, 2005: "Global Environmental Change: A Challenge for Human Security—Defining and Conceptualising the Environmental Dimension of Human Security", in: Rechkemmer, Andreas (Ed.): *UNEO—Towards an International Environment*

Organization—Approaches to a Sustainable Reform of Global Environmental Governance (Baden-Baden: Nomos): 85–109.

Bohle, Hans-Georg, 2002: "Land Degradation and Human Security", in: Plate, Erich (Ed.): *Environment and Human Security, Contributions to a Workshop in Bonn* (Bonn).

Brauch, Hans Günter, 2005: *Environment and Human Security. Freedom from Hazard Impact*, InterSecTions, 2/2005, UNU-EHS, Bonn.

Brauch, Hans Günter, 2008: "From a Security Towards a Survival Dilemma", in: Brauch, Hans Günter; Oswald Spring, Úrsula; Mesjasz, Czeslaw; Grin, John; Dunay, Pal; Behera, Navnita Chadha; Chourou, Béchir; Kameri-Mbote, Patricia, Liotta, P.H. (Eds.): *Globalization and Environmental Challenges: Reconceptualizing Security in the 21st Century* (Berlin–Heidelberg–New York: Springer-Verlag): 537–552.

Brauch, Hans Günter, 2009: "Introduction: Facing Global Environmental Change and Sectorialization of Security", in: Brauch, Hans Günter; Oswald Spring, Úrsula; Grin, John; Mesjasz, Czeslaw; Kameri-Mbote, Patricia, Behera, Navnita Chadha; Chourou, Béchir; Krumme-nacher, Heinz (Eds.), 2009: *Facing Global Environmental Change: Environmental, Human, Energy, Food, Health and Water Security Concepts* (Berlin–Heidelberg–New York: Springer-Verlag): 27–44.

Brauch, Hans Günter; Oswald Spring, Úrsula, 2009: *Securitizing the Ground, Grounding the Security* (Bonn: UNCCD).

Brauch, Hans Günter; Scheffran, Jürgen, 2012: "Introduction: Climate Change, Human Security and Violent Conflict in the Anthropocene", in: Scheffran, Jürgen; Brzoska, Michael; Brauch, Hans Günter; Link, Peter Michael; Schilling, Janpeter (Eds.): *Climate Change, Human Security and Violent Conflict: Challenges for Societal Stability* (Heidelberg: Springer).

Brooks, David, 2010: "Crece en México 'insurgencia' de cárteles: Clinton", in: *La Jornada*, 9 September 2010; at: http://www.jornada.unam.mx/2010/09/09/politica/002n1pol (17 May 2013).

Buzan, Barry; Waever, Ole; Wilde, Jaap de, 1998: *Security. A New Framework for Analysis* (Boulder, Co: Rienner).

CCI, 2012: "Fifth National Communication to the United Nations Framework Convention on Climate Change". *Executive summary* (México, D.F.: Comité Intersectorial sobre el Cambio Climático, SEMARNAT-INE).

CHS (Comisión on Human Security), 2003: *Human Security Now* (New York: UNO).

Coe, Michael D. (Ed.), ^{6}1999: *The Maya* (New York: Thames and Hudson).

Commission for Environmental Cooperation, s.d: "The North American Mosaic: An Overview of Key Environmental Issues"; at: http://www.cec.org/soe/files/en/SOE_landUse_en.pdf.

CONABIO-UAEM, 2006: Contreras, T.; Jaramillo, F.; Boyas, J.C. (Eds.): *La diversidad biológica en Morelos, Estudio del Estado* (México, D.F.: CONABIO-UAEM).

Conagua, 2008: *Programa Nacional Hídrico (PNH 2007–2012)* (México, D.F.: Conagua).

Conagua, 2014: *Programa Nacional Hídrico 2014–2018* (México, D.F.: Conagua).

Coneval, 2013: *Análisis y medición de la pobreza* (México, D.F.: Coneval).

Cook, Benjamin; Seager, Richard; Miller, Ron L., 2010: "Atmospheric Circulation Anomalies During Two Persistent North American Droughts: 1932–1939 and 1948–1957"; in: *Climate Dynamics*, 36: 2339–2355.

Cortés Muñoz, Juana Enriqueta; Martín Domínguez, Alejandra, 2012: "Disponibilidad de agua entubada y morbi-mortalidad por enfermedades infecciosas gastrointestinales en México", in: *XXII Congreso Nacional de Hidráulica,* Acapulco, Gro., November.

De la Torre González, Alejandra; Murillo Ramírez, Miguel Phalti, 2011: "La lucha contra el narco en México: desde una perspectiva de Seguridad Humana", in: *Observatório de Segurança Humana* (ISCSP-UTL).

Delgado, Gloria; Gutiérrez, Harim B., 2007: *Historia de México. De la era revolucionaria al sexenio de cambio*, 2 vol. (México, D.F.: Pearson Educación para México).

Dickie, A.; Streck, C.; Roe, S.; Zurek, M.; Haupt, F.; Dolginow, A., 2014: "Strategies for Mitigation Climate Change in Agriculture"; at: www.agriculturalmitigation.org.

Ensanut, 2012: *Encuesta Nacional de Salud y Nutrición. Resultados Nacionales* (Cuernavaca: INSP).
FAO, 1983: *World Food Security: A Reappraisal of the Concepts and Approaches.* Director General's Report (Rome: FAO).
FAO, 2010: *Biodiversity for Food and Agriculture. Contributing to Food Security and Sustainability in a Changing World* (Rome: FAO).
Forbes, 2014: "The World's Billioinaires", at: http://www.forbes.com/billionaires/list/#tab:overall.
Grin, John; Rotmanns, Jan; Schot, Johan, 2010: *Transitions to Sustainable Development. New Directions in the Study of Long Term Transformative Change* (New York: Routledge).
Huete, A.; Didan, K.; Miura, T.; Rodríguez, E.P.; Gao, X.; Ferreira, L.G., 2002: "Overview of Radiometric and Biophysical Performance of the MODIS Vegetation Indices", in: *Remote Sensing of Environment*, 83: 195–213.
IEA, 2014: http://www.iea.org/topics/energysecurity/ (January 2014).
INEGI, 1950–2000: *Censo General de Población y Vivienda* (Aguascalientes: INEGI).
INEGI, 2010: *Censo General de Población y Vivienda 2010* (Aguascalientes: INEGI).
INEGI, 2012: *Encuesta Nacional de Ingresos y Gastos de los Hogares 2010* (Aguascalientes: INEGI).
Informe Bourbaki, 2011: "El Costo Humano de la Guerra por la Construcción del Monopolio del Narcotráfico en México (2008–2009)", Universidad de Buenos Aires.
IPCC, 2012: *Special Report on Extreme Events (SREX)* (Geneva: IPCC).
IPCC, 2013: *Fifth Assessment Report: Climate Change 2013. The Physical Science Basis* (Geneva: IPCC).
IPCC, 2014: *Fifth Assessment Report: Climate Change 2014: Impacts, Adaptation, and Vulnerability* (Geneva: IPCC).
Jaso Galván, Azucena, 2013: "Guerra contra el narcotráfico: militarización y contrainsurgencia en México (2006–2012)", in: *Anais do V Simpósio Internacional Lutas Sociais na América Latina Revoluções nas Américas: passado, presente e futuro*, GEPAL, 10 a 13/09/2013, at: http://www.uel.br/grupo-pesquisa/gepal/v9_azucena_GIV.pdf.
Latinobarómetro, 2013: *Informe 2013, Latinobarómetro*; at: www.latinobarómetro.org.
Lenton, Timothy M.; Held, Hermann; Kriegler, Elmar; Hall, Jim W.; Lucht, Wolfgang; Rahmstorf, Stefan; Schellnhuber, Hans Joachim, 2008: "Tipping Elements in the Earth's Climate System", PNAS, 105,6 (12 February 12): 1786–1793.
Martín, Alejandra et al., 2011: "Assessment of a Water Utility Agency: A Multidisciplinary Approach", in: Oswald Spring, Úrsula (Ed.): *Water Resources in Mexico. Scarcity, Degradation, Stress, Conflicts, Management, and Policy* (Heidelberg: Springer-Verlag): 421–434.
Melillo, Jerry M.; Richmond, Terese; Yohe, Gary W. (Eds.), 2014: *Climate Change Impacts in the United States: The Third National Climate Assessment.* U.S. Global Change Research Program (Washington, D.C.: U.S. Government Printing Office).
Merkusheva, Natalia; Rapsomanikis, George, 2014: "Nonlinear Cointegration in the Food-Ethanol-Oil System. Evidence from Smooth Threshold Vector Error Correction Models", in: *ESA Working Paper* No. 14–01 (Rome: FAO).
Ministerial Declaration of The Hague, 2000: "Water Security" (The Hague: World Water Forum).
Ministry of Economy, 2012: *Analysis of the Corn-Tortilla Value Chain: Current Situation and Local Competition Factors* (Mexico D.F.: Ministry of Economy).
Morales, Novelo J.; Rodríguez, Tapia L., 2011: "The Growth of Water Demand in Mexico City and the Over-exploitation of its Aquifers", in: Oswald Spring, Úrsula (Ed.): *Water Resources in Mexico. Scarcity, Degradation, Stress, Conflicts, Management, and Policy* (Heidelberg: Springer-Verlag): 395–406.
MunichRe, 2008: *Topics Geo-Natural Catastrophes 2007. Analyses, Assessments, Positions* MunichRe, München; at: https://www.munichre.com/en/homepage/index.html.
Murphey, Paul, 2013: "Money Laundering and The Drug Trade: The Role of the Banks", in: *Global Research*, October.
OECD, 2011: *Revenue Statistics in Latin America 1990–2010* (Paris: OECD).

OECD, 2014a: *Economic Outlook, Analysis and Forecasts* (Paris: OECD).
OECD, 2014b: *Development Pathways Multidimensional* (Paris: OECD).
OIT [ILO], 2012: *Panorama Laboral de la OIT* (Geneva: ILO).
Oswald Spring, Úrsula, 2009a: "A HUGE Gender Security Approach: Towards Human, Gender and Environmental Security", in: Brauch, Hans Günter; Oswald Spring, Úrsula; Grin, John; Mesjasz, Czeslaw; Kameri-Mbote, Patricia; Behera, Navnita Chadha; Chourou, Béchir; Krumme-nacher, Heinz (Eds.): *Facing Global Environmental Change: Environmental, Human, Energy, Food, Health and Water Security Concepts* (Berlin–Heidelberg–New York: Springer-Verlag): 1165–1190.
Oswald Spring, Úrsula, 2009b: "Food as a New Human and Livelihood Security Challenge", in: Brauch, Hans Günter; Oswald Spring, Úrsula; Grin, John; Mesjasz, Czeslaw; Kameri-Mbote, Patricia; Behera, Navnita Chadha; Chourou, Béchir; Krumme-nacher, Heinz (Eds.): *Facing Global Environmental Change: Environmental, Human, Energy, Food, Health and Water Security Concepts* (Berlin–Heidelberg–New York: Springer-Verlag): 471–500.
Oswald Spring, Úrsula, 2011: *Water Resources in Mexico. Scarcity, Degradation, Stress, Conflicts, Management, and Policy* (Berlin–Heidelberg–New York: Springer-Verlag).
Oswald Spring, Úrsula, 2013: "Dual Vulnerability Among Female Household Heads", in: *Acta Colombiana de Psicología*, 16,2: 19–30.
Oswald Spring, Úrsula; Brauch, Hans Günter, 2009: "Water Security", in: Brauch, Hans Günter; Oswald Spring, Úrsula; Grin, John; Mesjasz, Czeslaw; Kameri-Mbote, Patricia; Behera, Navnita Chadha; Chourou, Béchir; Krumme-nacher, Heinz (Eds.): *Facing Global Environ-mental Change: Environmental, Human, Energy, Food, Health and Water Security Concepts* (Berlin–Heidelberg–New York: Springer-Verlag): 175–202.
Oswald Spring, Úrsula; Serrano Oswald, Serena Eréndira; Estrada Álvarez, Adriana; Flores Palacios, Fátima; Ríos Everardo, Maribel; Brauch, Hans Günter; Ruiz Pantoja, Teresita E.; Lemus Ramírez, Carlos; Estrada Villanueva, Ariana; Cruz Rivera M. Teresa Mónica, 2014: *Vulnerabilidad social y género entre migrantes ambientales* (Cuernavaca: CRIM-UNAM).
Pacheco, Julia; Cabra, Armando; Barcelo, Manuel; Alcocer, Ligia; Pacheco, Mercy, 2011: "Environmental Study on Cadmium in Groundwater in Yucatan", in: Oswald Spring, Úrsula (Ed.): *Water Resources in Mexico. Scarcity, Degradation, Stress, Conflicts, Management, and Policy* (Berlin–Heidelberg–New York: Springer-Verlag): 239–250.
Palacios Vélez, E.; Mejía Saez, E., 2011: "Water Use for Agriculture in Mexico", in: Oswald Spring, Úrsula (Ed.): *Water Resources in Mexico. Scarcity, Degradation, Stress, Conflicts, Management, and Policy* (Berlin–Heidelberg–New York: Springer-Verlag): 129–141.
Pérez Espejo, Rocío, 2011: "Contaminación del agua por la agricutura: retos de política y estudio de caso en Guanajuato", in: Oswald Spring, Úrsula (Ed.): *Water Resources in Mexico. Scarcity, Degradation, Stress, Conflicts, Management, and Policy* (Berlin–Heidelberg–New York: Springer-Verlag): 607–617.
Piñeyro, José Luis, 2012: "El ¿saldo? de la guerra de Calderón contra el narcotráfico", in: *El Cotidiano*, núm. 173: 5–13; at: http://www.elcotidianoenlinea.com.mx/pdf/17302.pdf.
PUED-UNAM (Programa Universitario de Estudios del Desarrollo), 2013: at: http://www.pued.unam.mx/index.php/pued.
Qiu, Cheng; Colson, Gregory; Escalante, Cesar; Wetzstein, Michael, 2012: "Considering Macroeconomic Indicators in the Food Before Fuel Nexus", in: *Energy Economics*, 34,6 (November): 2021–2028.
Rangel Medina, M.; Monreal Saavedra, R.; Watts, C., 2011: "Coastal Aquifers of Sonora: Hydrogeological Analysis Maintaining a Sustainable Equilibrium", in: Oswald Spring, Úrsula (Ed.): *Water Resources in Mexico. Scarcity, Degradation, Stress, Conflicts, Management, and Policy* (Berlin–Heidelberg–New York: Springer-Verlag): 65–73.
Robles Berlanga, Hector, 2010: "The Long-Term View: Comparing the Result of Mexico's 1991 and 2007 Agricultural Censuses", in: Fox, J.; Haight, L. (Eds.): *Subsidizing Inequality: Mexican Corn Policy Since NAFTA* (Washington, D.C.: Woodrow Wilson International Center for Scholars; Mexico, D.F.: Centro de Investigacion y Docencia Economicas, University of California, Santa Cruz).

Romm, Joe, 2011: at: ClimateProgess.org.

Rosengaus, M., 2007: "Informe interno", *Procedimientos para estimar tendencias del análisis parcial de datos históricos, 40 años de datos diarios, Tmax, Tmin y precipitación, tendencias, promedio nacional, todos los meses y anuales de 1961 a 2000 a nivel nacional, regional y estatal* (Mexico, D.F.: Coordinación General del Servicios Meteorológico).

Sánchez Cohen, Ignacio; Oswald Spring, Úrsula; Díaz, Gabriel, 2011: "Forced Migration by Climate Change in Mexico. Some Functional Relationships", in: *Journal for International Migration,* online.

Sener (Ministry of Energy), 2012: *Perspectivas de Petróleo Crudo 2012–2016* (México, D.F.: Sener).

Serra, T.; Zilberman, D.; Gil, J.M.; Goodwin, B.K., 2011: "Nonlinearities in Oil-Gasoline Price System", in: *Agricultural Economics*, 42: 35–45.

SHCP (Ministry of Finance), 2011: Data Bank; at: www.shcp.gob.mx.

SIPRI, 2013: *SIPRI Yearbook 2013. Armaments, Disarmament and International Security* (Oxford: Oxford University Press).

Stratfor, 2010: at: http://www.stratfor.com/topics/terrorism-and-security/drug-trafficking.

Turrent Fernández, Antonio; Wise, Timothy A.; Garvey, Elise, 2013: "Achieving Mexico's Maize Potential", in: *Intern. Conference* (New Haven: Yale University): September 14–15.

UNDP, 1994–2014: *Human Development Report* (Oxford: Oxford University Press).

USDA, 2013: *Crop Production Report.*

Wæver, Ole, 2008: "Peace and Security: Two Evolving Concepts and Their Changing Relationship", in: Brauch, Hans Günter; Oswald Spring, Úrsula; Mesjasz, Czeslaw; Grin, John; Dunay, Pal; Behera, Navnita Chadha; Chourou, Béchir; Kameri-Mbote, Patricia; Liotta, P.H. (Eds.): *Globalization and Environmental Challenges: Reconceptualizing Security in the 21st Century.* Hexagon Series on Human and Environmental Security and Peace, vol. 3 (Berlin–Heidelberg–New York: Springer-Verlag): 99–112.

WEF (World Economic Forum), 2013: *Water Security. The Water-Energy-Food-Climate Security Nexus* (Washington, D.C.: Island Press).

Wise, Timothy A., 2012: "The Impacts of U.S. Agricultural Policies on Mexican Producers", Paper 8, *Global Development and Environment Institute*, Tufts University; at: http://www.indexmundi.com/agriculture/?country=mx&commodity=corn&graph=imports.

Chapter 7
Building Sustainable Peace by Moving Towards Sustainability Transition

Hans Günter Brauch

Abstract This chapter focuses on the hypothetical implications of the uncertain outcomes of a long-term transformative change that will achieve sustainable development through a process of a sustainability transition. It addresses the question of whether a long-term transformative change might result in a more peaceful environment. The chapter is structured in ten parts. After a brief introduction, it discusses sustainable development as a goal and sustainability transition as a transformative process. It reviews the scientific debate on sustainability transition and its impact on the report *A Social Contract for Sustainability*, examines the climate and energy policy initiatives of the European Union, and analyses policy debates on climate and energy policy issues. The argument takes up the consequences of the human intervention in the earth system, with which we are threatening the survival of humankind. The sustainable 'peace concept' is briefly conceptualized for the Anthropocene; its realization requires major innovations in economic and environment policy. It points up contested visions, strategies and policies aiming at a sustainable peace with the goal of avoiding the security implications of climate change and countering resource conflicts, and it concludes with a discussion of the need to develop strategies and policies for sustainability transition that will lead to a 'sustainable peace' in the Anthropocene.

Keywords Sustainable development · Sustainability transition · Sustainability transition research network (STRN) · Energy transition · Low-carbon economy · Decarbonization · Sustainable peace · EU · G7 · BRICS

The author is grateful for critical comments and valuable suggestions to Juliet Bennett, University of Sydney, Australia; Dr. Carl Bruch, Washington DC, USA; Prof. Dr. Simon Dalby, Wilfrid Laurier University, Waterloo, Canada; Prof. Dr. Kalevi Holsti, University of British Columbia, Vancouver, Canada, and Prof. Dr. Ken Conca, American University, Washington DC, USA.

PD Dr. Hans Günter Brauch, chairman, Peace Research and European Security Studies (AFES-PRESS), Mosbach, Germany; email: brauch@afes-press.de.

H.G. Brauch et al. (eds.), *Addressing Global Environmental Challenges from a Peace Ecology Perspective*, The Anthropocene: Politik—Economics—Society—Science 4, DOI 10.1007/978-3-319-30990-3_7

7.1 Introduction

The Secretary-General of the United Nations, Ban Ki-Moon, in his report to the General Assembly on "Climate Change and its Possible Security Implications" of 11 September 2009 (A/64/350), distinguished between two debates on climate change and security by referring to climate change as a 'threat multiplier' for possible security threats at the community, national, regional and international levels, and as a 'threat minimizer' when aiming for sustainable development (UNSG 2009).

In previous publications, Brauch (2009, 2014) reviewed these two policy debates and the emerging scientific discourse. This chapter carries the argument further by discussing the realization of a 'sustainable transition' by contributing to the goal of a sustainable peace (Oswald Spring et al. 2014; Brauch et al. 2016; Brauch 2016). This chapter focuses on the hypothetical implications of the uncertain and unpredictable outcomes of a possible long-term transformative change that aims to realize of the goal of sustainable development through a process of sustainability transition that cannot be tested by scientific means, because the specific events that initiate structural changes in the production and consumption processes can be neither foreseen nor predicted.

The policy goal has been addressed in many policy declarations calling for a decarbonized world by the end of the twenty-first century, as in the *Leaders' Declaration* of the G7 Summit at Castle Elmau (Germany) on 7–8 June 2015:

> as a common vision for a global goal of greenhouse gas emissions reductions we support sharing … the latest IPCC recommendation of 40–70 % reductions by 2050 compared to 2010 recognizing that this challenge can only be met by a global response. We commit to doing our part to achieve a low-carbon global economy in the long term including developing and deploying innovative technologies striving for a transformation of the energy sectors by 2050 and invite all countries to join us in this endeavour. To this end we also commit to develop long-term national low-carbon strategies.[1]

To achieve these goals the heads of state and governments of the G7 countries and of the European Union committed themselves to a "mobilization of private sector capital … for … unlocking the required investments in low-carbon technologies as well as in … building resilience against the effects of climate change." With this, they promised to intensify their efforts (a) to increase the insurance for "up to 400 million … people in the most vulnerable developing countries … against the negative impact of climate change related hazards by 2020 and [to] support the development of early warning systems in the most vulnerable countries"; and (b) to "accelerate access to renewable energy in Africa and developing countries … with a view to reducing energy poverty and mobilizing substantial financial resources

[1]See the text of the final document at: http://www.consilium.europa.eu/en/meetings/international-summit/2015/06/7-8/.

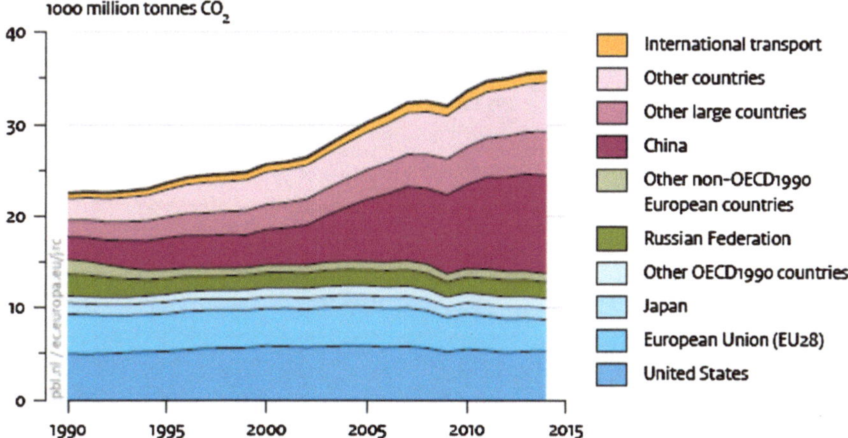

Fig. 7.1 Global CO_2 emissions per region from fossil-fuel use and cement production (1990–2014). *Source* Figure 2.1 in: Olivier et al. (2015), *Trends in global CO_2 emissions; 2015 Report* (The Hague: PBL Netherlands Environmental Assessment Agency; Ispra: European Commission, Joint Research Centre, 2015): 11; at: http://edgar.jrc.ec.europa.eu/news_docs/jrc-2015-trends-in-global-co2-emissions-2015-report-98184.pdf based on EDGAR 4.3 (JRC/PBL, 2015) (1970–2012; notably IEA 2014 and NBS 2015); EDGAR 4.3FT2014 (2013-2014): BP 2015; GGFR 2015; USGS 2015; WSA 2015. See there also in Figure 2.2: "CO_2 emissions from fossil-fuel use and cement production in the top 5 emitting countries and the EU"

from private investors, development finance institutions and multilateral development banks by 2020".

In the years ahead it will be seen whether these policy declarations by the G7 will become reality. Their implementation will face the powerful opposition of the hydrocarbon (coal, oil, gas) and nuclear lobbies and their ideological allies in the media, in special interest groups and in political parties, especially in the US, Canada (Dalby 2016), Australia and Japan, but also in many threshold and developing countries.

The rapidly developing regions as well as threshold and BRICS states have increased their CO_2 emissions from 1992 to 2010 exponentially, e.g. China by +240 %, India by +157 %, Brazil by +91 %, and South Africa by +45 % (see Fig. 2.10 above and Fig. 7.1; IEA 2014 and PBL/JRC 2015).

In 2010, the twelve G7 and BRICS countries[2] contributed nearly 70 %, the G20 about 80 % of global CO_2 emissions from energy consumption, while the remaining 175 countries contributed only about 20 % (Table 7.1).

This text addresses the question of why a new long-term transformative change may result in a more peaceful environment, while all previous long-term changes in human history have resulted in more deadly forms of warfare (as with both the Neolithic and Industrial Revolutions, see Brauch, Chap. 2 above). As such

[2]These countries are identified in Table 7.1.

Table 7.1 Carbon Dioxide Emissions from Energy Consumption (1990–2012)

Rank, 2010	Country	Rank, 1992	Rank, 2009	Change in rank, 1992 to 2010	% change in emissions, 1992 to 2010	Per person emissions, 2010, tonnes	2010
	World total				48	4.6	31,780.36
	North America				12	14.5	6,605.67
9	Canada	9	7	0	13	16.3	548.75
14	Mexico	14	13	0	42	4.0	445.28
2	United States	1	2	-1	10	18.1	5,610.11
13	Brazil	19	15	6	91	2.3	453.87
	Europe				1	7.2	4,370.29
17	France	11	18	-6	3	6.2	395.20
6	Germany	5	6	-1	-11	9.6	793.66
15	Italy	10	17	-5	0	7.2	416.37
24	Turkey	25	24	1	91	3.4	263.54
10	United Kingdom	7	10	-3	-8	8.5	532.44
	Eurasia				-24	8.7	2,454.13
4	Russia	3	4	-1	-19	11.7	1,633.80
	Middle East				119	8.4	1,785.93
47	Israel	57	46	10	68	9.6	70.32
51	Qatar	66	50	15	153	76.9	64.68
11	Saudi Arabia	20	12	9	103	18.6	478.41
	Africa				51	1.1	1,145.16
27	Egypt	36	27	9	110	2.4	196.55
12	South Africa	13	11	1	45	9.5	465.10
	Asia & Oceania				150	3.7	14,161.44
16	Australia	16	16	0	47	18.8	405.34
57	Bangladesh	80	57	23	244	0.4	56.74
1	China	2	1	1	240	6.3	8,320.96
3	India	6	3	3	157	1.4	1,695.62
18	Indonesia	23	14	5	116	1.6	389.43
5	Japan	4	5	-1	8	9.2	1,164.47
7	Korea, South	15	9	8	97	11.9	578.97
29	Malaysia	40	28	11	150	6.4	181.93
22	Thailand	33	23	11	176	4.2	278.49
37	Vietnam	77	38	40	549	1.3	112.80
G7							**9,461.00**
BRICS							**12,569.35**
12 (G7 and five BRICS) countries							**22,030.35**

Note The seven G7 countries are marked in yellow and the five BRICS countries in green.

Source US Energy Information Administration (US–EIA); *Guardian*, 21 June 2012 'structure-creating events' cannot be predicted and longer-term trends can only be projected, any answer will be tentative. The key thesis is that since the Industrial Revolution, human beings have directly interfered in the earth system by burning cheap fossil energy sources and thus have become both the 'cause' as well as the 'victims' of the consequences of global environmental and climate change.

During the twenty-first century the policy performance of these twelve G7 and BRICS countries will largely determine whether the vision of a decarbonized world may become a reality or whether dangerous climate change (Schellnhuber et al. 2006) with the possibility of climate-induced violent conflicts may become a reality.

The chapter is structured in ten parts. The next part discusses sustainable development as a goal and sustainability transition as a transformative process (7.2); it reviews the scientific debate on sustainability transition and its impact on the policy report *A Social Contract for Sustainability* (7.3); it examines the climate and energy policy initiatives of the European Union using a time perspective of up to 2030 and 2050 (7.4) and analyses the interaction of policy debates and scientific discourse on climate and energy policy issues (7.5). The argument then takes up the consequences of the human intervention in the earth system where "We are threatening the survival of humankind" (7.6), and the sustainable 'peace concept' is briefly conceptualized for the Anthropocene (7.7). This means that building sustainable peace requires major innovations not only in foreign and defence policy but also in economic and environment policy (7.8). The next part points to contested visions, strategies and policies aiming at a sustainable peace (7.9) with the goal of avoiding the security implications of climate change and countering resource conflicts, and the chapter concludes with a discussion of the need to develop strategies and policies for sustainability transition that will lead to a 'sustainable peace' in the Anthropocene (7.10).

7.2 Sustainable Development as a Goal and Sustainability Transition as a Long-Term Transformative Process

7.2.1 Sustainable Development in the UN Framework

Policies for a low-carbon economy aim to achieve sustainable development as defined by the Brundtland Report (WCED 1987), that is, a form of development that "meets the needs of the present without compromising the ability of future generations to meet their own needs".[3] This Report portrayed sustainable development as "a process of change in which the exploitation of resources, the direction of investments, the orientation of technological development, and institutional

[3]United Nations, 1987: "Report of the World Commission on Environment and Development" (New York: UN).

change are all in harmony and enhance both current and future potential to meet human needs and aspirations". The three pillars of sustainable development are economic growth, environmental protection and social equality.

In June 1992, the *United Nations Conference on Environment and Development* (UNCED) resulted in the signing of the *UN Framework Convention on Climate Change* (UNFCCC) and of the *UN Convention on Biodiversity* (CBD), the adoption of *Agenda 21* and a mandate for negotiating a *UN Convention to Combat Desertification* (UNCCD), signed in 1994. The *United Nations Commission on Sustainable Development* (UNCSD) was established by the *United Nations General Assembly* (UNGA) in December 1992 to ensure an effective follow-up to the *United Nations Conference on Environment and Development* (UNCED) in Rio.

In 2002 the *United Nations Conference on Sustainable Development* (UNCSD) reviewed achievements and shortcomings, and adopted the Johannesburg Declaration on Sustainable Development and a Plan of Implementation of the World Summit on Sustainable Development. In June 2012, as the outcome of the United Nations Conference on Sustainable Development (Rio+20), the conference approved a legally non-binding policy document on "The Future we Want",[4] which proposed a set of *Sustainable Development Goals* (SDGs), guidelines on green economic policies, and a ten-year framework for sustainable consumption and production. The SDGs, defined in the report of an intergovernmental committee of experts on sustainable development financing, and other documents were adopted at the UN Summit in September 2015 by heads of state and government and high representatives[5] in a document entitled "Transforming our world: the 2030 Agenda for Sustainable Development".

The 17 SDGs "with 169 associated targets" include a "transformational vision" of "a world in which every country enjoys … sustainable economic growth and decent work for all", where "consumption and production patterns and use of all natural resources … are sustainable". The SDGs are general policy guidelines for a process of sustainability transition and call for "access to affordable, reliable, sustainable and modern energy for all" (goal 7); "sustainable economic growth" (goal 8); "inclusive and sustainable industrialization and … innovation" (goal 9); "sustainable consumption and production patterns" (goal 12); and a "global partnership for sustainable development" (goal 17).

Goal 12, "sustainable consumption and production patterns", includes "implement the 10-year framework of programmes on sustainable consumption and

[4]See at: https://sustainabledevelopment.un.org/ and at: https://sustainabledevelopment.un.org/futurewewant.html (22 January 2015); see: "Post-2015 Development Agenda" with access to all adopted documents at: https://sustainabledevelopment.un.org/post2015.

[5]See at: https://sustainabledevelopment.un.org/rio20 (22 January 2015); see: "Post-2015 Development Agenda" with access to all adopted documents at: https://sustainabledevelopment.un.org/post2015 (20 August 2015), including: "Transforming our World: the 2030 Agenda for Sustainable Development".

production" (12.1), so that it will be possible by 2030 to achieve "the sustainable management and efficient use of natural resources" (12.2), "encourage companies … to adopt sustainable practices and to integrate sustainability information into their reporting cycle" (12.6), and "support developing countries to strengthen their scientific and technological capacity to move towards more sustainable patterns of consumption and production" (12a), and to "rationalize inefficient fossil-fuel subsidies that encourage wasteful consumption by removing market distortions" (12c).

In this document the heads of state and government and high representatives committed themselves to implementing these goals at the national, regional and global levels. Whether these intentions will be more successful than previous policy declarations and the legally binding Kyoto Protocol (1997) will not depend on 'declaratory politics' but on the will of the citizens and their elected governments to fight for the implementation of these goals against opposition by powerful oligarchies, their economic and political lobbies, and the ideologies that support policies of *business-as-usual* and assume that their representatives have the power to realize their interests and the technical means to save their 'way of life' irrespective of global costs.

In the social and policy sciences, discourse analyses of declaratory politics will remain at a surface level; the analysis of the social movements and non-governmental agencies in favour of a process of a sustainability transition may provide insights into the actors and their strategies, but policy-focused 'implementation and impact research' is needed to examine which strategic and technological innovations have succeeded or failed, and for which sociopolitical reasons. A research approach that examines sustainability transition may inspire both research and action in the development of both policy goals and implementation of the sustainability transition process.

7.2.2 A Dutch Research Project on Sustainability Transition

Unrelated to this UN debate, the theoretical and empirical debate on sustainability transition emerged from a research project of the *Dutch Knowledge Network on Systems Innovation and Transition* (KSI). Its theoretical results were published by Grin et al. (2010a, b) with, as its object, "to understand transitions dynamics, and how and to what extent they may be influenced". They were convinced

> that only through drastic system innovations and transitions it becomes possible to bring about a turn to a sustainable society to satisfy their own needs, as inevitable for solving a number of structural problems on our planet, such as the environment, the climate, the food supply, and the social and economic crisis.

They also argued that

> our world has to overcome the undesirable side effects of the ongoing 'modernization transition'.... However, the transition to sustainability has to compete with other developments, and it is uncertain which development will gain the upper hand. ... [They] ... closely address the need for transitions, as well as their dynamics and design (Grin et al. 2010a, b: xvii–xix).

The Dutch scholars, Geels and Schot (2010: 11–104), offered a sociotechnical and multilevel perspective on the dynamics of transitions; they introduced 'co-evolution' and 'multi-actor' processes, where radical shifts from one configuration to another one possibly materialize as part of "long-term processes" (forty to fifty years). They saw in Braudel's three historical times (see Brauch, Chap. 2) "useful general heuristics for studying long-term processes" involving "sociotechnical transitions". Geels (2002) distinguished in his multilevel perspective between a hierarchy of niches (of radical innovations), a patchwork of sociotechnical regimes, and the resulting new landscape or "exogenous context" (Fig. 7.2).

Their multilevel perspective was influenced by science and technology studies, evolutionary economics and sociology, including structuration and neo-institutional theory, and by the social mechanisms in agency–structure interactions. They offered a "typology of transition pathways", including (a) the "transformation pathway",

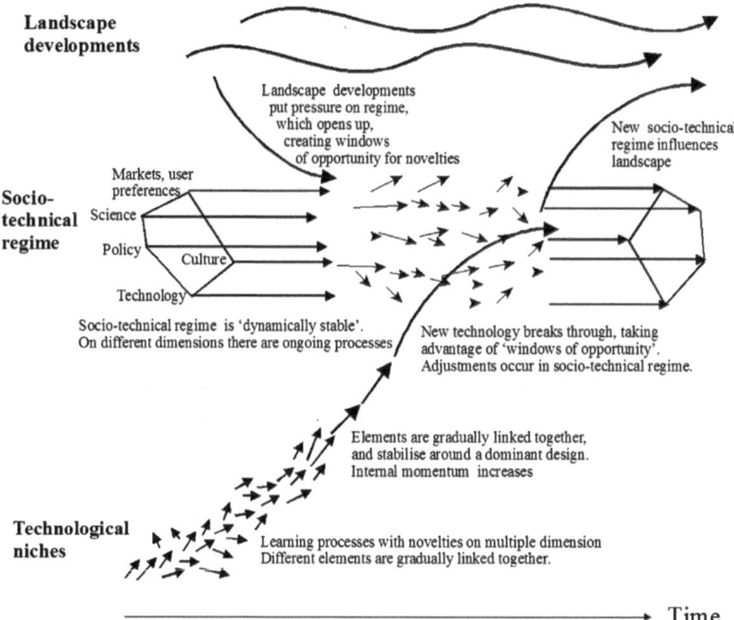

Fig. 7.2 Multilevel perspective on transitions. *Source* Geels and Schot (2010: 25), adapted from Geels (2002: 1263); used with the permission of the author Frank Geels

(b) the "de-alignment and re-alignment pathway", (c) the "technological substitution pathway", (d) a "reconfiguration pathway", and (e) "mixing pathways". They reviewed "managing sustainable innovation journeys" and in their causal analysis they focused on (i) outcomes, and (ii) the unfolding of processes, "by identifying patterns and underlying mechanism". While Rotmans/Loorbach (2010) proposed a systemic and reflexive approach for better understanding transitions, Grin (2010) tried to understand transition from a governance perspective influenced by the processes of institutional change and by an agency and analytical perspective. In their conclusions Grin, Rotmans and Schot discussed how to understand transitions and how to influence them.

These theoretically guided and empirically based case studies have inspired new directions in the study of long-term transformative change in the Netherlands and Flanders, and this initiative has spread to other parts of Europe since 2009 in the annual conferences of the *Sustainability Transition Research Network* (STRN), where social scientists from central and northern Europe predominated, with a few observers from North America, Australia and Asia.

7.2.3 The Sustainability Transition Research Network (STRN)

While both the UN and the European Union have developed the goal of sustainable development and outlined policy strategies for a transition towards a green and low-carbon economy, this policy debate has been detached from the scientific discourse on sustainability transition. Since 2009, the STRN[6] has focused from different scientific perspectives on 'sustainability problems' in the energy, transport, water and food sectors and on the ways

> in which society could combine economic and social development with the reduction of its pressure on the environment. A shared idea among these scholars is that due to the specific characteristics of the sustainability problems (ambiguous, complex) incremental change in prevailing systems will not suffice. There is a need for transformative change at the systems level, including major changes in production [and] consumption that were conceptualized as 'sustainability transitions'.[7]

The STRN has defined transitions research as

> a new approach to sustainable development (SD) … Major research efforts … have advanced knowledge of transitions to sustainability, particularly in the field of a broad understanding of how major, radical transformations unfold and what drives them. … Technical changes need to be seen in their institutional and social context, generating the notion of 'socio-technical (s-t) systems', which are often stable and path dependent, and

[6]See at: http://www.transitionsnetwork.org/ (22 January 2015).

[7]See at: http://www.transitionsnetwork.org/files/STRN_research_agenda_20_August_2010%282%29.pdf.

> therefore difficult to change. Under certain conditions and over time, the relationships within s-t systems can become reconfigured and replaced in a process that may be called a system innovation or a transition.

The STRN has argued that

> transitions to sustainability may turn out to be strongly context specific: dependent on the configurations of sectors and need areas, on national policy contexts and cultural aspects as well as on specific political contexts. It is therefore of great interest to explore the varied governance challenges that transitions to sustainability imply in different contexts. … from a variety of different research fields: industrial transformation, innovation and socio-technical transitions; integrated assessment; sustainability assessment; governance of SD (political science); policy appraisal community; researchers working on reflexive governance; the resilience community; the ecological economics community; groups of energy-, environment- and sustainability-modelers; and a core sustainability transitions community. …

According to the STRN's mission statement, the research of its participants

> is organized around seven themes: (a) synthesizing perspectives and approaches to transitions; (b) governance, power and politics; (c) implementation strategies; (d) civil society, culture and social movements in transitions; (d) firms and industry; (e) geography of transitions; (e) modelling of transitions.

The STRN has defined its mission as contributing "foresight reports on strategic sustainability policy questions … to support the development of a sustainability transitions research community internationally, and provide an independent, authoritative and credible source of analysis and insight into the dynamics and governance of sustainability transitions".[8]

Their multilevel perspective on transitions was influenced by Geels's (2002: 1263) model; this starts with many often interrelated and mutually reinforcing technological innovations in 'niches'. Such multiple events initially face opposition in the respective sociotechnical regime, and once they are overcome they often result in structural change by exploiting windows of opportunity and leading to changes in the landscape (Fig. 7.2).[9] Since its founding conference in Amsterdam (2009), the STRN has met in Lund (2011), Copenhagen (2012), Zürich (2013), Utrecht (2014) and Brighton (2015).[10] So far the STRN has remained an innovative 'niche' and has not become the mainstream of advanced sustainability studies.

---

[8]See the opening page of the STRN website, at: http://www.transitionsnetwork.org/ (22 January 2015).

[9]The *Environmental Innovation and Societal Transitions* (EIST) journal "offers a platform for reporting studies of innovations and socio-economic transitions to enhance an environmentally sustainable economy and thus solve structural resource scarcity and environmental problems, notably related to fossil energy use and climate change." See the EIST journal; at: http://www.journals.elsevier.com/environmental-innovation-and-societal-transitions/.

[10]See at: http://www.transitionsnetwork.org/events/conferences.

7.3 Towards a "Social Contract for Sustainability"

The Dutch Knowledge Network (KSI) and the international STRN influenced a policy report by the *German Advisory Council on Global Change* (WBGU 2011) on *A Social Contract for Sustainability* (2011), which argued that a low-carbon society needs to abolish subsidies for fossil fuel carriers and provide incentives for low-carbon enterprises. While climate protection is "a vital fundamental condition for sustainable development on a global level. …Sustainable development include [s] many other natural resources, such as fertile soil and biological diversity."

The WBGU (2011: 93) report adapted Geels's model and added several megatrends in both the earth system (climate, biodiversity, land degradation, water, raw materials) and the human system (development, democratization, energy, urbanization, food), where innovative changes in the regime may directly affect the megatrends (Fig. 7.3). The report argued that the realization of a low-carbon economy and society is overcoming the multiple barriers and exploiting the favourable factors (WBGU 2011: 6). The WBGU report stated that

> [t]he transformation into a sustainable society requires a modern framework to allow … almost nine billion people to lead 'the good life', both in terms of living with each other, and living with nature: a new Contrat Social … [that] represents a special agreement between science and society. … A low-carbon transformation can only be successful if it is a common goal, pursued simultaneously in many of the world's regions (WBGU 2011).

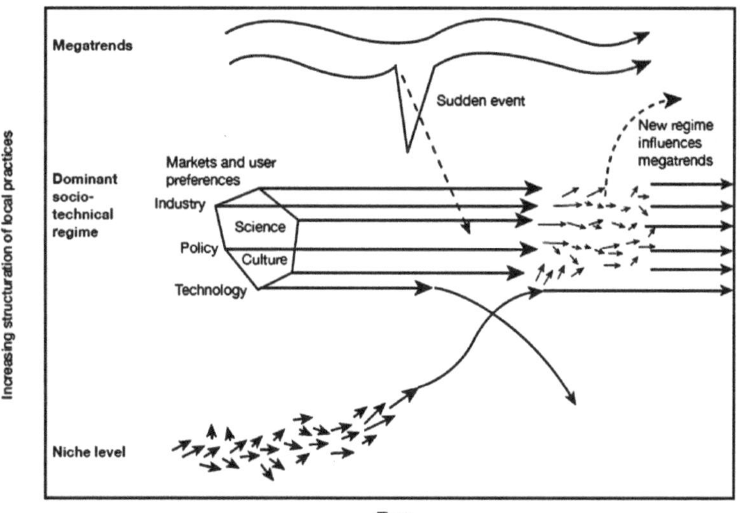

Fig. 7.3 Multilevel model for analysing transformation processes. *Source* WBGU (2011: 93), based on Grin et al. (2010a, b: 25, 69); adapted from Geels (2002: 1263). Reprinted with permission of WBGU

The WBGU (2011: 5) discussed the global "remodelling of economy and society towards sustainability as a 'Great Transformation'. Production, consumption patterns and lifestyles in all of the three key transformation fields must be changed in such a way that global greenhouse gas emissions are reduced to an absolute minimum over the coming decades, and low carbon societies can develop." The transformation towards a climate-friendly society requires that many existing change agents lead to a "concurrence of multiple change" (Osterhammel 2009, 2014), which "can trigger historic waves and comprehensive transformations".

The social dynamics for a change in the direction of climate protection must therefore be created through a combination of measures at different levels:

- It is knowledge based, based on a joint vision, and guided by the precautionary principle.
- It relies heavily on the change agents, who can test and advance the options for leaving behind an economy reliant on the use of fossil resources, thus helping to develop new leitmotifs, or new visions, to serve as guiding principles for social transition (Fig. 7.4).
- It needs a proactive state to allow the transformation process to develop into a certain direction by providing the relevant framework, by setting the course for structural change, and by guaranteeing the implementation of climate friendly innovations. The proactive state gives the change agents leeway, and supports them actively. It also counts on the cooperation of the international community and the establishment of global governance structures as the indispensable driving force for the intended transformation momentum (WBGU 2011: 5–6).

This transformation aims at a low-carbon society that starts with the decarbonization of energy systems, where "the proactive state and the change agents are the key players", who should jointly initiate within the next decade a transformation towards a sustainable energy path to allow a major decarbonization of the economy by 2050 while avoiding both a rebound effect and a climate crisis (WBGU 2011: 7).

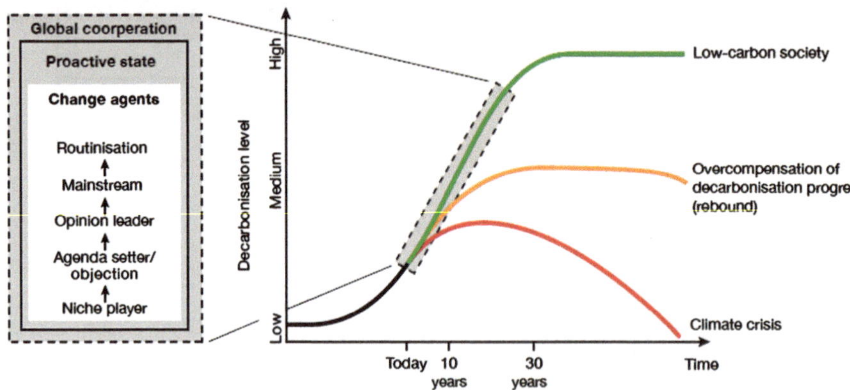

Fig. 7.4 The transformation's temporal dynamics and action levels. *Source* WBGU (2011: 7), modified according to Grin et al. (2010a, b). Reprinted with permission of WBGU

To achieve a major transformation towards a low-carbon economy and society, the WBGU proposed specific measures for the energy sector, changes in land use and global urbanization that could accelerate and extend the transition to sustainability.

1. The state should show conscious awareness of its enabling and proactive role to advance global decarbonization…
2. A European energy policy aiming for a fully decarbonized energy system by 2050 at the latest should be developed and implemented at once… One top priority for any development policy should be to provide access to sustainable energy to the 2.5 to 3 billion people in developing countries currently living in energy poverty.
3. A huge effort should be made to steer the world's accelerating urbanization towards sustainability.
4. Land-use can and should become climate-friendly, in particular forestry and agriculture.
5. Financing of the transformation and the massive investments required should increasingly rely on new business models that help to overcome current investment barriers.
6. Within international climate policy, states should continue to work towards an ambitious global treaty. At the same time, multilateral energy policy must promote the worldwide transfer of low-carbon technologies (WBGU 2011).

The WBGU Report proposed that "research and education are tasked with developing sustainable visions, in co-operation with policy-makers and citizens; identifying suitable development pathways, and realizing low-carbon and sustainable innovations". It suggested that

> during the establishment of low-carbon energy systems, the challenge lies in ending energy poverty in developing countries whilst also drastically, and quickly, mitigating global CO_2 emissions from the use of fossil energy carriers. … This requires efficiency improvements and lifestyle changes in many areas of people's everyday lives. Due to the high energy demand in cities, rapid urbanization is a central issue. From a technological point of view, there are various realistic options for the establishment of low-carbon energy systems. The WBGU recommends a strategy that relies primarily on an accelerated use of renewable energies. … The WBGU shows that transformation costs can be lowered significantly if joint decarbonization strategies are implemented in Europe. The transformation also represents a great chance for Europe to make innovation-driven contributions to a globalization process that has a viable future (WBGU 2011).

The report was discussed during an international symposium in Berlin on 9 May 2012, *Towards Low-Carbon Prosperity: National Strategies and International Partnerships*. Its three sessions focused on *Towards Low-Carbon Transformation*, *Sustainable Prosperity through Innovation* and *Pathways and Possibilities of Partnerships for Low-Carbon Prosperity*. In her remarks Chancellor Angela Merkel paired "climate change with efficient resource management or the problem of finite resources". She argued that "we are better off if we can dissolve our dependence on conventionally generated energy. The two crucial elements of the answer must therefore be changing our energy supplies, by switching to renewables, and dealing more efficiently with energy and the resources we have." She stressed that her government "decided to raise the proportion of renewables in our overall energy consumption to 60 % by 2050. For electricity consumption, that figure is to be 80 %."

Several reports by the *German Advisory Council on Global Change* (WBGU) have had a direct impact on the political agenda-setting brought about by the German government during her chairmanship of the EU (in 2007 on climate change and security: WBGU 2007, 2008; Brauch 2009, 2014), as well as on the G8 (Heiligendamm, 6–8 June 2007) and G7 (Elmau in 2015), where these goals were reflected in the final policy declarations.[11] In her remarks prior to Rio+20, Merkel also pointed to an ethical dimension of her vision for Germany and Europe:

> to conduct test phases, to learn how best to deal with the complex of new energy supplies, resource efficiency and efficient technology, and to subsidize progress. …We spent many years and decades overexploiting the world's resources. … We have a duty to redress the balance somewhat. I feel that we should step up to that duty and, what's more, turn it to our advantage. … The Green Economy Roadmap is of key importance, and we need to consolidate it at the United Nations. …

Chancellor Merkel encouraged the scientists to "stay stubborn" and told them "Don't be afraid to get on politicians' nerves from time to time. … Keep working to increase the community within our society of people who say yes, we need fundamental change." Thus, moving towards 'sustainability transition' is neither a pure scientific nor a technocratic *top-down* project, but requires the determined will and persuasive pressure of the citizens from the *bottom up* to change the values, preferences and behaviour of citizens, society, the business sector and the government. Merkel concluded that "sustainability needs to become a central tenet in every area of our lives", where it is necessary "to make the change we need happen. Convincing the majority is not always easy, but I believe it is our duty to do so."[12] In Europe, the EU and especially the European Commission and the Council, representing the twenty-eight EU member countries, have been a persistent source of forward-looking studies and policy analyses which have often been harshly attacked by the powerful economic lobbies.

---

[11]See the G8 Chair's summary, Heiligendamm, 8 June 2007: at: http://www.g-8.de/Content/EN/Artikel/__g8-summit/anlagen/chairs-summary,templateId=raw,property=publicationFile.pdf/chairs-summary.pdf: "In setting a global goal for emissions reductions in the process we have agreed in Heiligendamm … [on] at least a halving of global emissions by 2050. … Technology, energy efficiency and market mechanisms … are key to mastering climate change as well as enhancing energy security. … We agreed that energy efficiency and technology cooperation will be crucial elements of our follow-up dialogue." See the final declarations of the G7 summit in Elmau at: http://www.consilium.europa.eu/en/meetings/international-summit/2015/06/7-8/.

[12]See the text documentation from the symposium of 9 May 2012 in Berlin; at: http://www.wbgu.de/fileadmin/templates/dateien/symposium2012/Documentation_Symposium.pdf (23 August 2015).

7.4 Climate and Energy Policy Initiatives of the European Union up to 2030 and 2050

These policy proposals were partly taken up by the European Commission and the European Council in its longer-term goals and policy papers on climate change, its energy (EU 2010, 2011a, b), resource (Happaerts 2016; EU 2011c, d, e) and transport policies[13] (EU 2011f) and its "Roadmap for moving to a competitive low carbon economy in 2050" (EU 2011g).[14] In this Roadmap the European Commission addressed the goal "of reducing greenhouse gas emissions by 80–95 % by 2050 compared to 1990".

The *International Insitute for Applied Systems Analysis* (IASA) (with the EC4MACS consortium) provided quantitative technical analyses for the European Commission that would allow it "to develop scenarios until 2050 that would result in deep cuts in greenhouse gas emissions in the European Union. Scientists estimated the potentials for emission reductions from energy, agriculture and land use for CO_2 and the other greenhouse gases and calculated co-benefits of such low-carbon development paths on local air quality."[15]

The results of these analyses have been incorporated into: (a) the Communication of the Commission on "A Roadmap for moving to a competitive low carbon economy in 2050"[16]; (b) the Impact Assessment (SEC (2011) 288) of the European Commission, and (c) the Summary of the Impact Assessment. Based on these goals, the Commission's Roadmap outlined milestones with "policy challenges, investment needs and opportunities in different sectors". Based on modelling analysis of possible scenarios, the Roadmap concluded that

> domestic emission reductions of the order of 40 and 60 % below 1990 levels would be the cost-effective pathway by 2030 and 2040. … Such a pathway would result in annual reductions compared to 1990 of roughly 1 % in the first decade until 2020, 1.5 % in the second decade from 2020 until 2030, and 2 % in the last two decades until 2050. … Figure (7.5) illustrates the pathway towards an 80 % reduction by 2050… The upper 'reference' projection shows how domestic greenhouse gas emissions would develop under current policies. A scenario consistent with an 80 % domestic reduction then shows how overall and sectoral emissions could evolve, if additional policies are put in place, taking into account technological options (EU 2011h: 4).

[13]See "White paper 2011: Roadmap to a Single European Transport Area—Towards a competitive and resource efficient transport system"; at: http://ec.europa.eu/transport/themes/strategies/2011_white_paper_en.htm.

[14]European Commission, Climate Action, "Roadmap for moving to a low-carbon economy in 2050"; at: http://ec.europa.eu/clima/policies/strategies/2050/index_en.htm.

[15]See IASA: "The EU Roadmap for Moving to a Low Carbon Economy in 2050"; at: https://www.kowi.de/Portaldata/2/Resources/fp/Report-Towards-a-green-economy-in-Europe.pdf.

[16]See at: http://eur-lex.europa.eu/legal-content/EN/ALL/?uri=CELEX:52011DC0112 (23 August 2015).

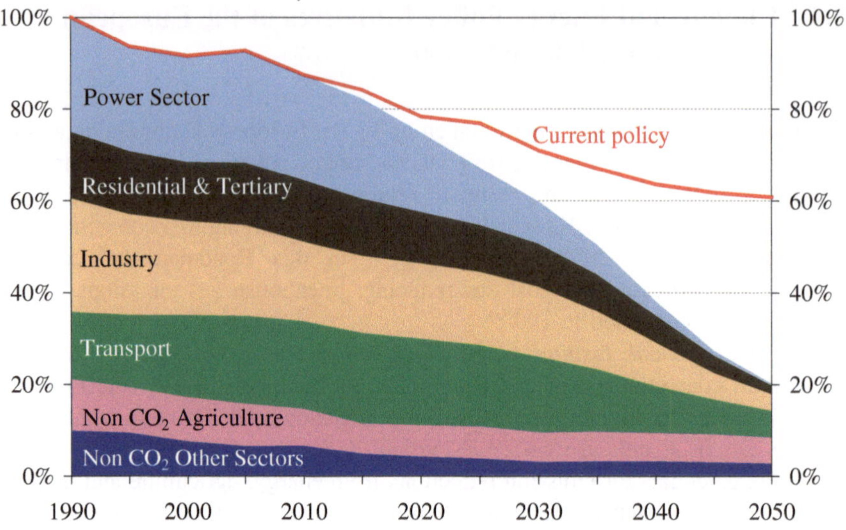

Fig. 7.5 EU GHG emissions towards an 80 % domestic reduction (100 % = 1990). *Source* EU (2011h: 5)

The Commission projected the GHG reductions needed in key sectors (Table 7.2). The modelling analysis assumed that "the switch to domestically produced low carbon energy sources will reduce the EU's average fuel costs by between €175 billion and €320 billion per year" (EU 2011h: 11). The study further assumed that "in 2050, the EU's total primary energy consumption could be about 30 % below 2005 levels" and it argued that "without action the oil and gas import bill could instead double compared to today, a difference of €400 billion or more per annum by 2050, the equivalent of 3 % of today's GDP" (EU 2011h: 12).

Table 7.2 Sectoral reductions

GHG reductions compared to 1990	2005 (%)	2030 (%)	2050 (%)
Total	−7	−40 to −44	−79 to −82
Sectors			
Power (CO_2)	−7	−54 to −68	−93 to −99
Industry (CO_2)	−20	−34 to −40	−83 to −87
Transport (incl. CO_2 aviation, excl. maritime)	+30	+20 to −9	−54 to −67
Residential and services (CO_2)	−12	−37 to −53	−88 to −91
Agriculture (non-CO_2)	−20	−36 to −37	−42 to −49
Other non-CO_2 emissions	−30	−72 to −73	−70 to −78

Source EU (2011h: 6)

The European Commission's Roadmap study for moving towards a low-carbon economy indicated

> that a cost effective and gradual transition would require a 40 % domestic reduction of greenhouse gas emissions compared to 1990 as a milestone for 2030, and 80 % for 2050. … With existing policies, the EU will achieve the goal of a 20 % GHG reduction domestically by 2020. … Deep reductions in the EU's emissions have the potential to deliver benefits in the form of savings on fossil fuel imports and improvements in air quality and public health. … The Roadmap gives ranges for emissions reductions for 2030 and 2050 for key sectors (EU 2011h: 14).

The summary of the Impact Asessment of the different EU decarbonization scenarios, reflecting the results of scientific modelling, concluded:

> that by 2050, an 80 % EU internal reduction compared to 1990 is technically feasible with proven technologies if a sufficiently strong carbon price incentive is applied across all sectors (range of around €100–€370 per ton of CO_2-eq. by 2050). This will require substantial continued innovation in existing technologies but is possible without the deployment of break-through technologies, such as nuclear fusion, hydrogen and fuel cells, or an electricity grid with widescale application of distributed energy storage, and without major lifestyle changes (e.g. dietary changes, strong changes in mobility patterns). Such developments could further facilitate a low carbon economy, but were not included in the analysis considering the uncertainties of their technical and economic feasibility and because of the difficulties of including them in the modelling tools.[17]

The Commission planned to use this Roadmap for sector-specific policy initiatives and longer-term funding considerations on how EU funding could support necessary instruments and investments for the transition to a low-carbon economy. Three years later, on 23 October 2014, the European Council agreed

- the domestic 2030 greenhouse gas reduction target of at least 40 % compared to 1990 together with the other main building blocks of the 2030 policy framework for climate and energy. … This 2030 policy framework aims to make the European Union's economy and energy system more competitive, secure and sustainable and also sets a target of at least 27 % for renewable energy and energy savings by 2030.
- The framework presented will drive continued progress towards a low-carbon economy. It aims to build a competitive and secure energy system that ensures affordable energy for all consumers, increases the security of the EU's energy supplies, reduces our dependence on energy imports and creates new opportunities for growth and jobs.
- The 2030 framework … also takes into account the longer term perspective set out by the Commission in 2011 in the Roadmap for moving to a competitive low carbon economy in 2050, the Energy Roadmap 2050 (EU 2011a) and the Transport White Paper (EU 2011b). These documents reflect the EU's goal of

[17] See "Summary of the Impact Assessment" (Brussels, 8.3.2011, SEC (2011) 289 final): 6.

> reducing greenhouse gas emissions by 80–95 % below 1990 levels by 2050 as part of the effort needed from developed countries as a group.[18]

These studies indicated that the goals addressed by the G8 (2007–2011), which were not repeated during the US presidency of the Obama administration in 2012 at Camp David,[19] and by the G7 (Elmau 2015) were taken up and studied by the European Commission.

In early July 2015, some twenty countries met at the sixth *Clean Energy Ministerial* (CEM) in Mérida, Mexico to discuss ways "to accelerate a global clean energy revolution" that would focus on "technology innovation and increasing market share" in order to reduce clean energy costs. Although they had different priorities, they all agreed on "the importance of supporting the rapid growth of our global clean energy economy, [i.e.] … energy efficiency and a diversity of renewable resources such as solar, wind, hydro, sustainable biomass, and geothermal". Representing "90 % of global clean energy investment", these countries and the European Commission aimed at a more ambitious stage of "CEM 2.0."

> India became the first country in the world to comprehensively set quality and performance standards for super-efficient LED lighting, potentially avoiding the equivalent of 90 coal-fired power plants of emissions. … The Solutions Center helped Caribbean countries set an ambitious sustainable energy target of 47 % for 2027 that will help reduce their dependence on expensive oil-fired electricity generation.

At the Mérida Ministerial the representatives launched a "*Global Lighting Challenge* … to collectively reach cumulative sales of 10 billion high-efficiency, high-quality, affordable advanced lighting products. … The enabler … is the technology innovation that has lowered LED costs by a factor of 10 in just a few years." They also set up "a new *Power System Challenge* that will help us toward the clean, efficient, and reliable electricity grids of the future and to increased access". They announced that they would "significantly scale-up the *Clean Energy Solutions Center* with a wider network of technical advisors and with a new Finance Portal to provide access to the world's best clean energy finance expertise". The final CEM document was signed by

- Wan Gang, Minister of Science and Technology, People's Republic of China
- Miguel Arias Cañete, Commissioner for Climate Action and Energy, European Commission
- Ségolène Royal, Minister of Ecology, Sustainable Development and Energy, France
- Piyush Goyal, Minister of State for Power, Coal and New & Renewable Energy, India

[18]See: "2030 framework for climate and energy policies" (22 January 2015).

[19]See: "Camp David Declaration", Camp David, Maryland, United States, 18–19 May 2012; at: https://www.whitehouse.gov/the-press-office/2012/05/19/camp-david-declaration.

- Pedro Joaquín Coldwell, Secretary of Energy, Mexico
- Suhail Mohammed Al Mazrouei, Minister of Energy, United Arab Emirates
- Ernest Moniz, Secretary of Energy, United States.

The two largest GHG emitters, the US (2016) and China (2017), offered to host the next CEM Ministerials. According to an article in the *Times of India* of 31 July 2015, CEM "complements the international climate change discussions by serving as a premier forum to efficiently help each other achieve our respective clean energy goals". The article noted that

> We are in the midst of a global clean energy revolution. Amidst China's newly installed capacity of 94 million kilowatts in 2013, about 60 % came from non-fossil energy sources. India has announced an ambitious target to scale-up its renewable energy capacity from 30 GW presently to 175 GW by 2022. Mexico in 2015 reached 22.8 % of its power generation from clean energy technologies, and has set a target of 35 % by 2024. The European Union has reduced primary energy consumption in 2013 by 15.5 % compared to 2020 projections and with full implementation and monitoring of already-adopted energy efficiency legislation can achieve its 20 % energy efficiency target in 2020.

By end of 2015, at COP 21 in Paris, it will become evident whether a majority of state parties will be willing to make legally binding commitments to reduce their GHG emissions in the decades to come and even more to fully implement these commitments by moving gradually towards a low-carbon economy, as well as realizing a sustainability transition in the energy and production sectors and adopting national policies that support sustainable consumption by their citizens (Brauch et al. 2016). The process of moving towards a "sustainable energy transition" is increasingly driven not by climate change concerns but by economic incentives, as renewables have become competitive over the past twenty-five years.

7.5 Interaction of Policy Debates and Scientific Discourses

The scientific discourse and policy debates have interacted closely. The policy debate since the publication of the Brundtland Report (1987) has to some extent triggered funding for new scientific institutions and research projects. The scientific debate, meanwhile, has moved on from the need to develop an approach to zero growth towards a reduction in the overexploitation of nature and towards allowing the ecosystem services that are essential for humans and nature to recuperate. In the global public and policy debate there has been an overemphasis on GHG emissions, while the major destruction of biodiversity and negative impacts on ecosystems have often been ignored.

A 'climate paradox' (Brauch 2012) has emerged among some G8 countries, especially Canada, the US and Japan. Every year from 2007 to 2011, these countries declared their intention to reduce their GHG by 80 % by 2050. At the same time, they failed to achieve their modest commitments under the UNFCCC and the Kyoto Protocol by 2000 and by 2012; there was a lack of political will and ability to implement long-term policy statements coupled with a readiness to postpone tough decisions to their successors and the next generation.

This weak performance and implementation of goals in the reduction of GHG emissions will most severely affect the most socially and environmentally vulnerable developing and least developed countries. These have already seen the highest number of deaths and affected people. Economic losses due to climate-induced hazards were the highest in developed countries because of insurance. It is projected that many countries with continuing high population growth and a high level of people below the poverty line will also have a low level of resilience and limited capabilities for adaptation and mitigation during the twenty-first century (Fig. 7.6).

While the EU, UN, UNEP and OCED have suggested a transition towards a green economy, so far only a few countries have announced and initiated detailed policy programmes aimed at a sustainable energy transition (e.g. Germany,[20] the UK,[21] France,[22] the US,[23] Brazil[24] and China[25]). The fossil and nuclear energy industries and the car and highway lobbies have attacked the IPCC, which they have identified as a key messenger, by supporting a campaign by climate critics; in the US Congress they succeeded in blocking all climate bills put forward by the Obama administration.[26] It remains to be seen whether Obama's Clean Power Plan of August 2015 will overcome the opposition. The tar sands lobby in Canada (Dalby 2016) and the fracking industry in the US have invested heavily in new fossil technologies and have supported policies to prevent the implementation of

[20]For "Energy transition in Germany", see at: http://en.wikipedia.org/wiki/Energy_transition_in_Germany (25 January 2015), with additional updated sources.

[21]HM Government: *The UK Low Carbon Transition Plan—National strategy for climate and energy* (London: HM Government, 15 July 2009).

[22]France's new energy law of July 2015 plans to cut the nuclear share of electricity from 75 to 50 % by 2025, while "energy consumption is to be slashed 20 % from 2012 levels by 2030, with renewables increasing to 32 % of the mix"; see at: http://www.rtcc.org/2015/07/23/france-moves-away-from-nuclear-with-clean-energy-law/. (23 August 2015).

[23]On 3 August 2015, President Barack Obama announced that his "Clean Power Plan is to cut greenhouse gas emissions from US power stations by nearly a third within 15 years" and "to cut carbon emissions from the power sector by 32 % by 2030, compared with 2005 levels"; at: http://www.bbc.com/news/world-us-canada-33753067; see "Fact Sheet: President Obama to Announce Historic Carbon Pollution Standards for Power Plants"; at: https://www.whitehouse.gov/the-press-office/2015/08/03/fact-sheet-president-obama-announce-historic-carbon-pollution-standards (23 August 2015).

[24]See "Brazil backs long term zero carbon goal as Merkel visits"; at: http://www.climatechangenews.com/2015/08/21/brazil-backs-long-term-zero-carbon-goal-as-merkel-visits/. In August 2015, during a visit by chancellor Angela Merkel to Brazil, president Dilma Rousseff supported decarbonization of the global economy by 2100, thus backing the G7's long-term goal to phase out fossil fuels.

[25]China's report to the UNFCCC.

[26]Naomi Klein: "Capitalism vs. the Climate—Denialists are dead wrong about the science. But they understand something the left still doesn't get about the revolutionary meaning of climate change", in: *The Nation*, 28 November 2011; at: http://www.thenation.com/article/164497/capitalism-vs-climate (25 January 2014).

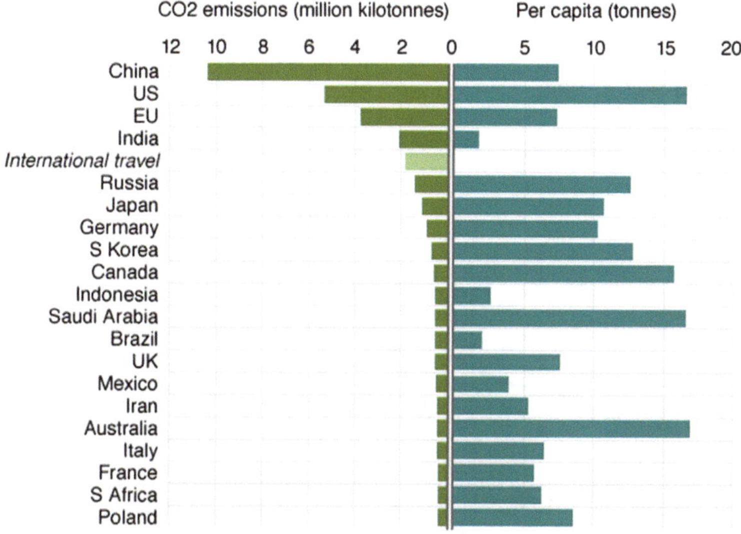

Fig. 7.6 Estimated 2013 CO_2 emissions in kilotonnes and per capita for the G20. *Source* BBC; at: http://www.bbc.com/news/world-us-canada-33753067 based on the EDGAR database of the European Commission and Netherlands Environmental Assessment Agency

legal obligations by not ratifying the Kyoto Protocol (US in 1998) or by withdrawing from it (Canada in 2011).

By 2013, China had reached a per capita CO_2 emission equivalent to the average of the twenty-eight EU countries and was producing nearly double the total CO_2 emissions of the US (Fig. 7.6). On 30 June 2015, China announced in its report to the UNFCCC secretariat that it aims (Fig. 7.7)

> to cut its greenhouse gas emissions per unit of gross domestic product by 60–65 % from 2005 levels. … China said it would increase the share of non-fossil fuels as part of its primary energy consumption to about 20 % by 2030. China plans to increase its installed capacity of wind power to 200 GW and solar power to around 100 gigawatts (GW), up from 95.81 and 28 GW today, respectively. It will also increase its use of natural gas which is expected to make up more than 10 % of its primary energy consumption by 2020. … China's plan will see it install as much low-carbon energy as the entire US electricity system capacity to date. Coal consumption still accounts for around 66 % of China's energy consumption. [In 2014] China's cabinet announced a plan to cap coal consumption by 2020 at a level of 4.2 bn tonnes and for coal to make up no more than 62 % of the primary energy mix by the same year.[27]

[27]See: "China makes carbon pledge ahead of Paris climate change summit", in: *The Guardian*, 30 June 2015; at: http://www.theguardian.com/environment/2015/jun/30/china-carbon-emissions-2030-premier-li-keqiang-un-paris-climate-change-summit.

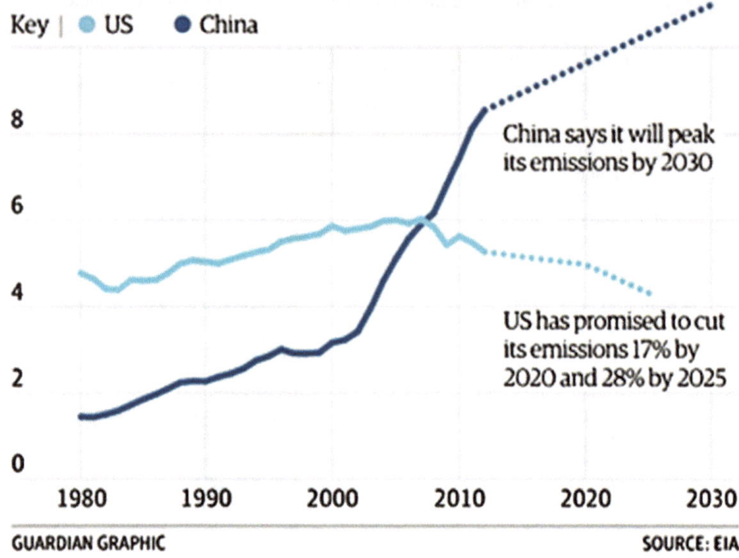

Fig. 7.7 CO_2 emissions of the USA and China in billion tonnes since 1980 and China's stated policies to curb their increase by 2030. *Source* "China makes carbon pledge ahead of Paris climate change summit", in: *The Guardian*, 30 June 2015, based on US EIA; at: http://www.theguardian.com/environment/2015/jun/30/china-carbon-emissions-2030-premier-li-keqiang-un-paris-climate-change-summit

According to China's report of 30 June 2015 to the UNFCCC Secretariat, it listed among "Policies and Measures to Implement Enhanced Actions on Climate Change" by 2030:

> A. Implementing Proactive National Strategies on Climate Change; … B. Improving Regional Strategies on Climate Change; … C. Building Low-Carbon Energy System; … D. Building Energy Efficient and Low-Carbon Industrial System; … E. Controlling Emissions from Building and Transportation Sectors; … F. Increasing Carbon Sinks; … G. Promoting the Low-Carbon Way of Life; … H. Enhancing Overall Climate Resilience; … I. Innovating Low-Carbon Development Growth Pattern; … J. Enhancing Support in terms of Science and Technology; … K, Increasing Financial and Policy Support; … L. Promoting Carbon Emission Trading Market; … M. Improving Statistical and Accounting System for GHG Emissions; … N. Broad Participation of Stakeholders. O. Promoting International Cooperation on Climate Change.[28]

Will these strategies, policies and measures, if they should be fully implemented in the decades ahead, have any impact on the conceptual discussion on peace and

[28]See China's report to the UNFCCC; at: http://www4.unfccc.int/submissions/INDC/Published%20Documents/China/1/China's%20INDC%20-%20on%2030%20June%202015.pdf.

security? Will such policy initiatives pose new economic challenges and security threats or might a transition process contribute to a 'sustainable peace'?

Possible 'peace dividends' of a sustainability transition in the energy sector, by increasing energy efficiency and switching gradually to renewables, may reduce demand for and dependence on coal, oil and gas imports. Whether this would reduce the probability and intensity of future resource conflicts cannot be foreseen and can only be speculated about. Strategies, policies and measures aiming at a sustainability transition in key economic sectors will require a combination of unilateral national and regional steps, as suggested by the European Commission, and lasting international agreements within multilateral frameworks in the economic, energy, and environmental sectors.

In 2015, policy discussions on a decoupling of economic growth from fossil energy consumption (UNEP 2011) and on a green economy (OECD)[29] continued and long-term policy documents by the European Commission (EU 2010, 2011d; Happaerts 2016) were being pursued. At the G7 meeting in Elmau (Germany) the heads of major industrialized countries stated:

> We affirm our strong determination to adopt at the Climate Change Conference in December in Paris this year (COP 21) a protocol, another legal instrument or an agreed outcome with legal force under the United Nations Framework Convention on Climate Change (UNFCCC). … This should enable all countries to follow a low-carbon and resilient development pathway in line with the global goal to hold the increase in global average temperature below 2 °C. … In order to incentivize investments towards low-carbon growth opportunities we commit to the long-term objective of applying effective policies and actions throughout the global economy.[30]

Whether the policy goals and legally nonbinding obligations that were adopted in the Paris Agrement in December 2015 will be fully implemented will be seen in the years to come. Readers will be able to judge whether COP 21 in Paris will initiate and reinforce a policy transition towards a decarbonization of the economy or whether short-term economic and political interests in the framework of *business-as-usual* will prevail. So far, theoretical, empirical and conceptual debates on sustainability transition (WBGU 2011) have had little influence on political agenda-setting and actual policy implementation.

7.6 We Are Threatening the Survival of Humankind

In mainstream thinking on security (Brauch et al. 2008, 2009, 2011), it is the 'other' who poses essential challenges and threats to 'us' as individuals, as an ethnic or religious community, as a state or country, or as a military alliance. Proponents of a

[29]See publications on: "Green growth and sustainable development", at: http://www.oecd.org/greengrowth/.

[30]See "G7 Leaders' Declaration, Schloss Elmau, Germany, June 8, 2015"; at: https://www.whitehouse.gov/the-press-office/2015/06/08/g-7-leaders-declaration (14 August 2015).

securitization of climate change argue that in the Anthropocene, 'humankind' has become *the* threat, or "we are the threat and we are the victims but the 'we' are not identical".

However, the interest of policymakers in the climate change and security nexus has differed with regard to national, international, and human security, as has the assessment by social scientists of the linkage between climate change and conflict (Burke et al. 2009; Buhaug 2010; Buhaug et al. 2014; Hsiang et al. 2013; Theisen et al. 2013; Gleditsch 2015; Salehyan 2014; Salehyan/Hendrix 2014; Ide et al. 2016):

- From the perspective of *international security*, many UN member states have emphasized in the General Assembly (UNGA 2009) and in the Security Council (UN 2007, 2011; EU 2008a; b) the need to strengthen sustainability policies and measures in order to prevent climate change becoming a 'threat multiplier' that may trigger a violent escalation of existing conflicts, and thus to minimize security threats. A major focus has been on preventing conflicts from escalating into violence when triggered by the physical and societal impacts of climate change.
- From the perspective of *US national security*, the interest of the defence and intelligence community is in how the US military can operate in a world where climate change impacts are increasing, and how the US can maintain its position as the single military superpower and influence outcomes in the interest of its national security. Thus, the focus is on conflict management but also on prevention (NIC 2008, 2012).
- From the perspective of *human security*, the goal has been to avoid climate-induced violent conflicts occurring that would affect the livelihood of human beings, especially of those with the highest social vulnerability in the poorest countries who lack the capacity for proactive adaptation and mitigation and whose capacity for resilience is limited (Brauch/Scheffran 2012; IPCC 2014a).

The cause of this 'new' climate-change-induced threat is no longer the military posture or behaviour of an adversary, but our own economic behaviour, or the increase in the burning of hydrocarbons (coal, oil and natural gas) in the Anthropocene due to modern production and consumption processes. Historically, the contribution of the industrialized countries to the accumulation of GHG in the atmosphere was more significant, but this is rapidly changing in China where GHG emissions per capita have reached and already overtaken that of some industrialized countries (e.g. of Italy and France, Fig. 7.5). But those most affected by extreme natural events are countries in the tropics and especially the poorest, who are the most vulnerable. This poses equity problems that have been raised by developing countries, who have called for financial transfers within the UNFCCC framework (IPCC 2012, 2014a, b).

Thus, policies that aim at a 'sustainability transition' in the energy (and other sectors) and that move towards a low-carbon economy may counter the security consequences of climate change impacts in a framework of *business-as-usual* and in a world where *Hobbesian* military options and *cornucopian* strategies aiming at geoengineering dominate.

Strategies, policies and measures aiming at a 'sustainability transition' may reduce the conflict potential and bring about peace dividends and possibly contribute to the 'utopian' vision of a sustainable peace. This is at present a pure heuristic question that needs the attention of experts in environmental, development and sustainability research as well as in peace and security studies.

7.7 'Sustainable Peace': Challenges of the Anthropocene

The policy debates and scientific discourses on 'sustainability transition' have excluded possible linkages between a fundamental macro-structural change or a new sociotechnical revolution and considerations of international peace and security. In their introduction to *Expanding Peace Ecology*, Oswald Spring et al. (2014) explored the two parallel conceptual debates on peace and ecology and the linkages between the five key concepts of peace, security, sustainability, equity, and gender. They argued (Fig. 7.8):

> While both the scientific peace and ecology concepts have significantly changed since the end of the Cold War, the scientific exchange between peace research and the different ecological approaches has been limited and most research occurred within the confinements of the respective research programmes. The conceptual bridge-building by Kenneth and Elise Boulding since the 1960s had few followers, while the policy debate and scientific research on the linkages between environment or ecology and security rapidly expanded (Oswald Spring et al. 2014: 14).

Oswald Spring, Brauch and Tidball suggested conceptualizing 'peace ecology' in the Anthropocene

> within the framework of five conceptual pillars … consisting of peace, security, equity, sustainability and gender. To conceptualize the linkages between peace and security we refer to 'negative peace' and for the relationship between peace and equity we use the

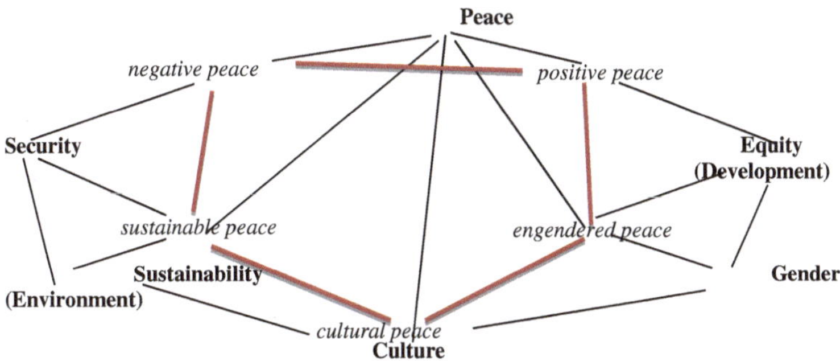

Fig. 7.8 Five Pillars of Peace Ecology and their four linkage concepts of negative, positive, cultural and engendered peace. *Source* Oswald Spring et al. (2014: 19)

> 'positive peace' concept, for interactions between peace, gender and environment we suggest the 'cultural peace' concept and finally for the relations between peace, equity and gender we propose the concept of an 'engendered peace' (Oswald Spring et al. 2014: 18).

They further argued that *sustainable peace* refers

> to the manifold links among peace, security and the environment, where humankind and environment as two interdependent parts of global Earth face the consequences of destruction, extraction and pollution. The sustainable peace concept includes also processes of recovering from environmental destruction, reducing human footprint in ecosystems through less carbon-intensive, and in the long-term possibly carbon-free and increasingly dematerialized production processes, so that future generations may still be able to decide on their own resources & development strategies (Oswald Spring et al. 2014: 18).

In the framework of the emerging discussion on "sustainability transition and sustainable peace in the Anthropocene" this author tried to carry the conceptual debate on the 'sustainable peace' concept further (Brauch 2016).

It was argued above (Chap. 2) that in the Anthropocene with the intervention of humankind in the earth system and in nature 'we are the threat' that is posed by 'our' economic behaviour and by the prevailing production, transportation and consumption processes that rely heavily on the burning of hydrocarbons for the production of our food (agriculture), goods (industry) and services (communication), and for our movement (by car, train, ship or plane). If 'we are the threat', then a sustainable peace policy in the Anthropocene has to address these multiple causes that are totally unrelated to classic security or military considerations. Such a sustainable peace policy must address the obstacles of policies aiming at a sustainability transition in major economic sectors, but also in the demand side. This is influenced by our values, perceptions, world views and mindsets, all of which in turn influence our economic, societal, political and consumptive behaviour.

A sustainable peace policy requires fundamental changes in our agricultural, economic, housing, transportation, and environmental policies, which should aim at a gradual decarbonization of energy and other key sectors that have been the major producers of greenhouse gases, and most particularly of CO_2. This means challenging the dominant Hobbesian thinking that aims at power-based military solutions to control those countries that own the largest reserves of fossil energy sources, most particularly the petrol and natural gas in Russia, North Africa, the Middle East, Nigeria, Venezuela, Mexico and others (Klare 2001, 2012, 2013).

7.8 Contested Visions, Strategies and Policies

Oswald Spring and Brauch (2011: 1487) argued that in the Anthropocene era of earth and human history (Crutzen/Stoermer 200; Crutzen 2002, 2011; Steffen et al. 2011) humankind is confronted with opposite visions of the future: *business-as-usual* in a Hobbesian world where economic and strategic interests and behaviour dominate and will lead to a major crisis for humankind that puts the survival of the vulnerable at risk; and the need for a *transformation* of global cultural, environmental, economic and political relations. The two visions address different strategies for coping with *global environmental change* (GEC):

- In the first vision of *business-as-usual* and *cornucopian* perspectives prevail that suggest primarily technical fixes ..., defence of economic, strategic and national interests with adaptation strategies that are in the interest of and affordable for ... OECD countries. ...
- In the alternative vision of a comprehensive transformation a *sustainable perspective* has to be developed and implemented into effective new strategies and policies with different goals and means based on global equity and social justice (Oswald Spring/Brauch 2011: 1487).

The possible consequences of the first vision are an increase in the probability of chaotic GEC and climate change (Schellnhuber et al. 2006), with both linear and chaotic changes in the climate system and their sociopolitical consequences, while the second vision "requires a change in *culture* (thinking on the human–nature interface), *worldviews* (thinking on the systems of rule, e.g. democracy vs. autocracy and on domestic priorities and policies as well as on interstate relations in the world), *mindsets* (strategic perspectives of policy-makers) and new forms of national and global *governance*" (Oswald Spring/Brauch 2011: 1487–1488).

This alternative vision addresses the need for a "new paradigm for global sustainability" (Clark et al. 2004), for a "transition to [a] much more sustainable global society" (Raskin et al. 2002), aimed at peace, freedom, material well-being and environmental health. Changes in technology and management systems alone will not be sufficient; "significant changes in governance, institutions and value systems" (Steffen et al. 2004: 291–293) are needed.

This sceptical diagnosis addresses two different approaches to international security and environmental policy. While adherents of a *business-as-usual policy* argue that the market, economic initiatives and military power will be able to cope with its consequences, the proponents of the alternative perspective emphasize the need for multiple efforts to move towards a long-term transition towards sustainability and to start with the decoupling of economic growth from an increase in fossil energy consumption and GHG emissions (UNEP 2014; von Weizsäcker 2014).

7.9 Avoiding the Security Implications of Climate Change and Countering Resource Conflicts

Instead of a 'militarization' of the possible socio-economic and security implications of an anthropogenic global climate change throughout this century, this author, influenced by Wæver (1995, 1997), has argued that a successful 'securitization' of climate change requires 'extraordinary measures' that address the causes.

Such a securitization has so far failed in the aftermath of COP 15 of UNFCCC in Copenhagen in December 2009.

While the structural trends of population growth can be projected, and different macroeconomic performance, including GHG emissions, can be modelled, the security consequences of reactive or proactive policy decisions to deal with both the linear and chaotic consequences of the climate system cannot be predicted (Gaddis 1992/1993). It is argued here that an alternative vision that aims for a sustainability transition together with concrete policies that will result in a gradual decarbonization of the economy may be the best way to avoid the possible negative security consequences of global anthropogenic climate change.

Since 1972, several reports to the Club of Rome on *The Limits to Growth* (Meadows et al. 1972, 1992, 2004; Randers 2012; Marino 2016) have pointed to major global resource constraints that may lead to resource conflicts. This perspective has been heavily criticized as a 'neo-Malthusian' approach by mainstream economists, many of them adherents of a 'cornucopian approach' (Gleditsch 2003), who argue that technological innovations have overcome resource constraints in the past.

Contrary to the concept of 'peak oil', *cornucopian* critics have pointed to new unforeseen oil reserves and to alternative fossil energy sources (tar sands, fracking of natural gas) and to new technologies for substituting other substances for oil (algae; conversion of seawater into jet fuel, or alternative biofuels, batteries, fuel cells, hydrogen etc.). While some authors have referred to the end of the 'Oil Age' (Leggett 2001, 2005a, b), others (e.g. Lovins et al. 2005) have suggested *Winning the Oil Endgame* by large-scale energy innovation, thus creating jobs and profits and maintaining energy security.

In 1981, a study by the Trilateral Commission on the *Security of the West* called for a Western military intervention capability that would protect access to oil resources and their transportation, which were essential in order for the modern economies of highly industrialized countries to function (Kaiser et al. 1981a, b). In the document *Defense Planning Guidance, FY* [Fiscal Years] *1994–1999*, deputies to Paul Wolfowitz, Under Secretary of Defense for Policy in the Bush administration, called in 1992 for a military capability identical to that of the cold war:

> to prevent the re-emergence of a new rival, either on the territory of the former Soviet Union or elsewhere, that poses a threat on the order of that posed by the former Soviet Union. This is a dominant consideration underlying the new regional defense strategy and requires that we endeavour to prevent any hostile power from dominating a region whose resources would, under consolidated control, be sufficient to generate global power. These regions include Western Europe, East Asia, the territory of the former Soviet Union, and Southwest Asia (Klare 1995: 101).[31]

A gradual decoupling from reliance on fossil energy supplies by oil-rich countries may remove a major cause of Western military interventions in oil-rich Muslim countries in the Middle East. Thus a decarbonization of western European

[31]This quote is based on a report in the *New York Times* of 8 March 1992.

countries may have a military and political pay-off—a reduction of the probability of energy-induced interventions in that region.

While powerful interest groups in the primary exporting countries of coal, oil (including from tar sands) and natural gas (including from fracking) have countered the debates on global climate change and the call for a decarbonization of the economy, net importing countries of fossil energy sources (e.g. the EU) are increasingly relying on renewable energy sources (wind, solar energy) for electricity. While both new fossil energy sources and renewables may counter resource conflicts, only the latter may avoid the negative security consequences of the physical effects of anthropogenic climate change (Klare 2001).

As a 'decoupling of growth from energy consumption' (UNEP 2014; von Weizsäcker 2014) is possible with an energy efficiency improvement of a factor 4, 5 or 10 (von Weizsäcker et al. 1997, 2009; von Weizsäcker 2014) by the replacement of fossil with renewable energy sources, dependence on energy imports will gradually decline and resource (oil) wars may thus also decline. This has also been a stated goal of the EU's Roadmap for a low-carbon economy by 2050 (EU 2011h).

7.10 Strategies and Policies for Sustainability Transition for 'Sustainable Peace' in the Anthropocene

Since World War II, in the social sciences, peace research and environmental studies have developed separately and there has been only very limited debate between the representatives of the two research communities (Stephenson 2016). This has also been the case with the narrower debates on the security consequences of climate change (Gleditsch 2012, Scheffran et al. 2012) and on transitions to sustainable development (Grin et al. 2010a, b). In a handbook on *Sustainability Transition and Sustainable Peace*, Brauch et al. (2016) offer conceptual, theoretical and empirical analyses that try to link both scientific discourses, as the UN Secretary-General Ban Ki-Moon has suggested in his report on Climate Change and Security (UNSG 2009).

This chapter has argued the case why a long-term transformative change towards sustainability in the framework of a low-carbon economy and society may result in a more peaceful environment, while all previous long-term changes in human history have resulted in deadlier forms of warfare. As has been argued above (in Chap. 2), such necessary 'structure-creating events' cannot be predicted, but current longer-term trends can be projected, which is why any answer must remain tentative. As human beings have directly interfered in the earth system since the Industrial Revolution by burning cheap fossil energy sources and thus have become the 'cause' but also the 'victims' of the consequences of global environmental and climate change, we as part of the human species can also become part of the solution if we are to be part of the "change we want the world to see" (Gandhi). However, this requires major changes in our own values, preferences and consumptive behaviour, and such changes require alternative pathways to achieve sustainable sectoral policies during this century that can drastically reduce our carbon footprint and help realize the goal of a low-carbon economy.

Among social scientists there is a need to overcome professionalization through overspecialization and to enter into a dialogue between environmental studies and peace research. Those authors who have proposed the concepts of "Spaceship Earth" (Boulding 1966, 1970), 'ecodynamics' (Boulding 1978, 1983), "environmental peacemaking" (Conca 1994; Conca/Dabelko 2002) and "peace ecology" (Kyrou 2007; Oswald Spring et al. 2014; Amster 2014) have suggested that such conceptual bridge-building is needed in order to understand the complexity of the linkages between sustainability and peace issues.

The scientific debate on 'sustainability transition' addresses the numerous scientific, societal, economic, political and cultural needs to reduce GHG emissions, and not only by legally binding quantitative emission reduction obligations. These have not achieved their goals during the past two decades because of a lack of political will and capability to implement these obligations.

References

Amster, Randall, 2014: *Peace Ecology* (Boulder, CO: Paradigm).

Boulding, Kenneth E., 1966: "The Economics of the Coming Spaceship Earth", in: Jarrett, H. (Ed.): *Environmental Quality in a Growing Economy* (Baltimore: Johns Hopkins Press).

Boulding, Kenneth E., 1970: "The Economics of the Coming Spaceship Earth", in: Boulding, Kenneth E., (Ed.): *Beyond Economics: Essays on Society, Religion, and Ethics* (Ann Arbor, University of Michigan Press): 275–287.

Boulding, Kenneth E., 1978: *Ecodynamics* (London: Sage).

Boulding, Kenneth E., 1983: "Ecodynamcis", in: *Interdisciplinary Science Reviews,* 8,2 (June): 108–113.

Brauch, Hans Günter, 2009: "Securitzing Global Environmental Change", in: Brauch, Hans Günter; Oswald Spring, Úrsula; Grin, John; Mesjasz, Czeslaw; Kameri-Mbote, Patricia; Behera, Navnita Chadha; Chourou, Béchir; Krummenacher, Heinz (Eds.): *Facing Global Environmental Change: Environmental, Human, Energy, Food, Health and Water Security Concepts* (Berlin–Heidelberg–New York: Springer-Verlag): 65–102.

Brauch, Hans Günter, 2012: "Climate Paradox of the G-8: Legal Obligations, Policy Declarations and Implementation Gap", in: *Revista Brasileira de Politica International*. Special issue edited by Eduardo Viola & Antonio Carlos Lessa on: *Global Climate Governance and Transition to a Low Carbon Economy* (Brasilia: Instituto Brasileira de Realacoes Internacionais): 30–52.

Brauch, Hans Günter, 2014: "From Climate Change and Security Impacts to Sustainability Transition: Two Policy Debates and Scientific Discourses", in: Úrsula Oswald Spring, Hans Günter Brauch, Keith G. Tidball (Eds.): *Expanding Peace Ecology: Peace, Security, Sustainability, Equity and Gender: Perspectives of IPRA's Ecology and Peace Commission* (Heidelberg–Dordrecht–London–New York: Springer-Verlag): 33–61.

Brauch, Hans Günter, 2016: "Conceptualizing Sustainable Peace in the Anthropocene: A Challenge and Task for an Emerging Political Geoecology and Peace Ecology", in: Brauch, Hans Günter; Oswald Spring, Úrsula; Grin, John; Scheffran; Jürgen (Eds.): *Handbook on Sustainability Transition and Sustainable Peace* (Cham–Heidelberg–New York–Dordrecht–London: Springer International Publishing).

Brauch, Hans Günter; Scheffran, Jürgen, 2012: "Introduction", in: Scheffran, Jürgen; Brzoska, Michael; Brauch, Hans Günter; Link, Peter Michael; Schilling, Janpeter (Eds.): *Climate Change, Human Security and Violent Conflict: Challenges for Societal Stability* (Berlin–Heidelberg–New York: Springer): 3–40.

Brauch, Hans Günter; Oswald Spring, Úrsula; Mesjasz, Czeslaw; Grin, John; Dunay, Pal; Behera, Navnita Chadha; Chourou, Béchir; Kameri-Mbote, Patricia; Liotta, P. H. (Eds.), 2008: *Globalization and Environmental Challenges: Reconceptualizing Security in the 21st Century* (Berlin–Heidelberg–New York: Springer-Verlag).

Brauch, Hans Günter; Oswald Spring, Úrsula; Grin, John; Mesjasz, Czeslaw; Kameri-Mbote, Patricia; Behera, Navnita Chadha; Chourou, Béchir; Krumme-nacher, Heinz (Eds.), 2009: *Facing Global Environmental Change: Environmental, Human, Energy, Food, Health and Water Security Concepts* (Berlin–Heidelberg–New York: Springer-Verlag).

Brauch, Hans Günter; Oswald Spring, Úrsula; Mesjasz, Czeslaw; Grin, John; Kameri-Mbote, Patricia; Chourou, Béchir; Dunay, Pal; Birkmann, Jörn (Eds.), 2011: *Coping with Global Environmental Change, Disasters and Security—Threats, Challenges, Vulnerabilities and Risks* (Berlin–Heidelberg–New York: Springer-Verlag).

Brauch, Hans Günter; Oswald Spring, Úrsula; Grin, John; Scheffran; Jürgen (Eds.), 2016: *Handbook on Sustainability Transition and Sustainable Peace* (Cham–Heidelberg–New York–Dordrecht–London: Springer International Publishing).

Brundtland Commission (World Commission on Environment and Development), 1987: *Our Common Future. The World Commission on Environment and Development* (Oxford–New York: Oxford University Press).

Buhaug, Halvard, 2010: "Climate not to blame for African civil wars", in: *PNAS*, 107,38: 16477–16482.

Buhaug, Halvard; Nordkvelle, Jonas; Bernauer, Thomas Bernauer; Böhmelt, Tobias; Brzoska, Michael; Busby, Joshua W.; Ciccone, Antonio; Fjelde, Hanne; Gartzke, Erik; Gleditsch, Nils Petter; Goldtsone, Jack A.; Hegre, Håvard; Holtermann, Helge; Koubi, Vally; Link, P. Michael; Link, Jasmin S. A.; Lujala, Päivi; O'Loughlin, John; Raleigh, Clionadh; Scheffran, Jürgen; Schilling, Janpeter; Smith, Todd G.; Theisen, Ole Magnus; Tol, Richard S. J.; Urdal, Henrik; von Uexkull, Nina, 2014: "One Effect to Rule Them All? A Comment on Climate and Conflict", in: *Climatic Change*, 127,3–4: 391–397.

Burke, Marshall B.; Miguel, Edward; Satyanath, Shanker; Dykema, John A.; Lobell, David B., 2009: "Warming Increases the Risk of Civil War in Africa", in: *PNAS*, 106,49: 20670–20674.

Clark, William C.; Crutzen, Paul J.; Schellnhuber, Hans Joachim, 2004: "Science and Global Sustainability: Toward a New Paradigm", in: Schellnhuber, Hans Joachim; Crutzen, Paul J.; Clark, William C.; Claussen, Martin; Held, Hermann (Eds.): *Earth System Analysis for Sustainability* (Cambridge, MA–London: MIT Press): 1–28.

Conca, Ken, 1994: "In the Name of Sustainability: Peace Studies and Environmental Discourse", in: *Peace and Change*, 19: 91–113.

Conca, Ken; Dabelko, Geoffrey D. (Eds.): *Environmental Peacemaking* (Baltimore: Johns Hopkins University Press–Woodrow Wilson Center Press).

Crutzen, Paul J., 2002: "Geology of Mankind", in: *Nature*, 415,3 (January): 23.

Crutzen, Paul J., 2011: "The Anthropocene: a geology of mankind", in: Brauch, Hans Günter; Oswald Spring, Úrsula; Mesjasz, Czeslaw; Grin, John; Kameri-Mbote, Patricia; Chourou, Béchir; Dunay, Pal; Birkmann, Jörn (Eds.): *Coping with Global Environmental Change, Disasters and Security—Threats, Challenges, Vulnerabilities and Risks* (Berlin–Heidelberg–New York: Springer-Verlag): 3–4.

Crutzen, Paul J.; Stoermer, Eugene F., 2000: "The Anthropocene", in: *IGBP Newsletter*, 41: 17–18.

Dalby, Simon, 2016: "Geopolitics, Ecology and Stephen Harper's Reinvention of Canada", in: Brauch, Hans Günter; Oswald Spring, Úrsula; Grin, John; Scheffran; Jürgen (Eds.): *Handbook on Sustainability Transition and Sustainable Peace* (Cham–Heidelberg–New York–Dordrecht–London: Springer International Publishing).

EU (European Commission), 2010: *Roadmap to a Resource Efficient Europe (Impact Assessment Roadmap)* (Brussels: European Commission).

EU (European Commission), 2011a: *Roadmap to a Resource Efficient Europe* (Brussels: European Commission).

EU (European Commission), 2011b: *Energy Roadmap 2050* (Brussels: European Commission).

EU (European Commission), 2011c: *Analysis Associated with the Roadmap to a Resource Efficient Europe. Part I* (Brussels: European Commission).

EU (European Commission), 2011d: *A resource-Efficient Europe—Flagship Initiative Under the Europe 2020 Strategy* (Brussels: European Commission).

EU (European Commission), 2011e: *Analysis Associated with the Roadmap to a Resource Efficient Europe. Part II* (Brussels: European Commission).

EU (European Commission), 2011f: *White Paper. Roadmap to a Single European Transport Area —Towards a Competitive and Resource Efficient Transport System* (Brussels: European Commission).

EU (European Commission), 2011g: *A Roadmap for Moving to a Competitive Low Carbon Economy in 2050* (Brussels: European Commission).

EU (European Commission), 2011h: Communication from the Commission to The European Parliament, The Council, The European Economic and Social Committee and The Committee of the Regions: *A Roadmap for Moving to a Competitive Low Carbon Economy in 2050.* COM (2011) 112 final (Brussels: European Commission, 8.3.2011).

EU (European Commission; Council), 2008a: *Climate Change and International Security.* Doc 7249/08 (Brussels: European Commission, 14 March).

EU (European Council), 2008b: *Report on the Implementation of the European Security Strategy —Providing Security in a Changing World*, S407/08 (Brussels: European Council, 11 December).

Gaddis, John Lewis, 1992/1993: "International Relations Theory and the End of the Cold War", in: *International Security*, 17,3 (Winter): 5–58.

Geels, Frank W., 2002: "Technological Transitions as Evolutionary Reconfiguration Processes: A Multi-level Perspective and a Case Study", in: *Research Policy*, 31,8–9: 1257–1274.

Geels, Frank W.; Schot, Johan, 2010: "The Dynamics of Transitions. A Socio-Technical Perspective", in: Grin, John; Rotmans, Jan; Schot, Johan (Eds.): *Transitions to Sustainable Development. New Directions in the Study of Long Term Transformative Change* (New York: Routledge): 11–104.

Gleditsch, Nils Petter, 2003: "Environmental Conflict: Neomalthusians vs. Cornucopians", in: Brauch, Hans Günter; Liotta, P. H; Marquina, Antonio; Rogers, Paul; Selim, Mohammed El-Sayed (Eds.): *Security and Environment in the Mediterranean. Conceptualising Security and Environmental Conflicts* (Berlin–Heidelberg: Springer): 477–486.

Gleditsch, Nils-Petter, 2012: "Whither the Weather? Climate Change and Conflict", in: *Journal of Peace Research, Special Issue: Climate Change and Conflict,* 49,1 (January–February): 9–18.

Gleditsch, Nils Petter, 2015: *Nils Petter Gleditsch: Pioneer in the Analysis of War and Peace* (Cham–Heidelberg–New York–Dordrecht–London: Springer).

Grin, John, 2010: "Understanding Transitions from a Governance Perspective", in: Grin, John; Rotmans, Jan; Schot, Johan: *Transitions to Sustainable Development. New Directions in the Study of Long Term Transformative Change* (New York, NY–London: Routledge).

Grin, John; Rotmans, Jan; Schot, Johan, 2010a: *Transitions to Sustainable Development. New Directions in the Study of Long Term Transformative Change* (New York, NY–London: Routledge).

Grin, John; Rotmans, Jan; Schot, Johan, 2010b: "The Dynamics of Transitions: A Socio-Technological Perspective", in: Grin, John; Rotmans, Jan; Schot, Johan (Eds.): *Transitions to Sustainable Development. New Directions in the Study of Long Term Transformative Change* (New York, NY–London: Routledge): 11–104.

Happaerts, Sander, 2016: "Discourse and Practice of Transitions in International Policy-making on Resource Efficiency in the EU", in: Brauch, Hans Günter; Oswald Spring, Úrsula; Grin, John;

Scheffran; Jürgen (Eds.): *Handbook on Sustainability Transition and Sustainable Peace* (Cham–Heidelberg–New York–Dordrecht–London: Springer International Publishing).

Hsiang, Solomon; Burke, Marshall; Miguel, Edward, 2013: "Quantifying the Influence of Climate on Human Conflict", in: *Science*, 341, 6151: 1–14.

Ide, Tobias; Link, P. Michael; Scheffran, Jürgen; Schilling, Janpeter, 2016: "The Climate–Conflict Nexus: Pathways, Regional Links, and Case Studies", in: Brauch, Hans Günter; Oswald Spring, Úrsula; Grin, John; Scheffran; Jürgen (Eds.): *Handbook on Sustainability Transition and Sustainable Peace* (Cham–Heidelberg–New York–Dordrecht–London: Springer International Publishing).

IPCC, 2011: *Renewable Energy Sources and Climate Change Mitigation* [SRREN] (Geneva: IPCC).

IPCC, 2012: *Managing the Risks of Extreme Events and Disasters to Advance Climate Change Adaptation* [SREX] (Geneva: IPCC).

IPCC, 2013: *Climate Change 2013—The Physical Science Basis. Working Group I Contribution to the Fifth Assessment Report of the Intergovernmental Panel on Climate Change* (Cambridge–New York: Cambridge University Press); at: http://www.ipcc.ch/report/ar5/wg1/.

IPCC, 2014a: *Climate Change 2014—Synthesis Report, Summary for Policymakers* (Geneva: IPCC); at: https://www.ipcc.ch/report/ar5/syr/.

IPCC, 2014b: "Human Security", in: *Climate Change 2014—Impacts, Adaptation, and Vulnerability—Part A: Global and Sectoral Aspects. Working Group II Contribution to the Fifth Assessment Report of the Intergovernmental Panel on Climate Change* (Cambridge–New York: Cambridge University Press): 755–701; at: http://www.ipcc.ch/pdf/assessment-report/ar5/wg2/WGIIAR5-Chap12_FINAL.pdf.

Kaiser, Karl; Lord, Winston; de Montbrial, Thierry; Watt, David, 1981a: *Die Sicherheit des Westens: Neue Dimensionen und Aufgaben* (Bonn: Deutsche Gesellschaft für Auswärtige Politik).

Kaiser, Karl; Lord, Winston; de Montbrial, Thierry; Watt, David, 1981b: *Western Security—What Has Changed? What Should be Done?* (New York: Council on Foreign Relations–London: Royal Institute of International Affairs).

Klare, Michael T., 1995: *Rogue States and Nuclear Outlaws. America's Search for a New Foreign Policy* (New York: Hill and Wang).

Klare, Michael T., 2001: *Resource Wars: The New Landscape of Global Conflict* (New York: Henry Holt–Metropolitan Books).

Klare, Michael T., 2012: *The Race for What's Left: The Global Scramble for the World's Last Resources* (London: Picador).

Klare, Michael T., 2013: "How Resource Scarcity and Climate Change Could Produce a Global Explosion", in: *The Nation*, 22 April); at: http://www.thenation.com/article/173967/how-resource-scarcity-and-climate-change-could-produce-global-explosion.

Kyrou, Christos N., 2007: "Peace Ecology: An Emerging Paradigm in Peace Studies", in: *The International Journal of Peace Studies*, 12,2 (Spring/Summer): 73–92.

Leggett Jeremy K., 2001: *The Carbon War: Global Warming and the End of the Oil Era* (London: Routledge).

Leggett Jeremy K., 2005a: *The Empty Tank: Oil, Gas, Hot Air, and the Coming Financial Catastrophe* (New York: Random House).

Leggett, Jeremy K., 2005b: *Half Gone: Oil, Gas, Hot Air and the Global Energy Crisis* (London: Portobello Books).

Lovins, Amory B.; Datta, E. Kyle; Bustness, Odd-Even; Koomey, Jonathan G.; Glasgow, Nathan J., 2005: *Winning the Oil Endgame: Innovation for Profit, Jobs and Security* (Snowmass: Rocky Mountain Institute).

Marino, Marit Sjøvaag, 2016: "From The Limits to Growth to 2052", in: Brauch, Hans Günter; Oswald Spring, Úrsula; Grin, John; Scheffran; Jürgen (Eds.): *Handbook on Sustainability Transition and Sustainable Peace* (Cham–Heidelberg–New York–Dordrecht–London: Springer, *forthcoming*).

Meadows, Donella H.; Meadows, Dennis L.; Randers, Jorgen; Behrens III, William W., 1972: *The Limits to Growth* (New York: Universe Books).

Meadows, Donella H.; Meadows, Dennis L.; Randers, Jorgen, 1982: *Beyond the Limits* (White River Jct., VT: Chelsea Green Publishing).

Meadows, Donella H.; Meadows, Dennis L.; Randers, Jorgen, 2004: *The Limits to Growth—the 30-Year Update* (White River Jct., VT: Chelsea Green Publishing).

NIC [National Intelligence Council], 2008: *Global Trends 2025: A Transformed World* (Washington DC: National Intelligence Council, November 2008).

NIC [National Intelligence Council], 2012: *Global Trends 2030: Alternative Worlds* (Washington DC: National Intelligence Council, 10 December 2012).

Nordås, Ragnhild; Gleditsch, Nils Petter, 2009: "IPCC and the climate–conflict nexus". Paper presented to the Synthesis Conference of the Global Environmental Change and Security program, Oslo, 22–24 June.

Olivier, Jos G.J.; Janssens-Maenhout, Greet; Muntean, Marilena; Peters, Jeroen A.H.W., 2015: *Trends in global CO_2 emissions; 2015 Report* (The Hague: PBL Netherlands Environmental Assessment Agency; Ispra: European Commission, Joint Research Centre, 2015).

Osterhammel, Jürgen, 2009: *Die Verwandlung der Welt. Eine Geschichte des 19. Jahr-hunderts* (München: C. H. Beck).

Osterhammel, Jürgen, 2014: *The Transformation of the World: A Global History of the Nineteenth Century* (Princeton: Princeton University Press).

Oswald Spring, Úrsula; Brauch, Hans Günter, 2011: "Coping with Global Environmental Change —Sustainability Revolution and Sustainable Peace", in: Brauch, Hans Günter; Oswald Spring, Úrsula; Mesjasz, Czeslaw; Grin, John; Kameri-Mbote, Patricia; Chourou, Béchir; Dunay, Pal; Birkmann, Jörn (Eds.): *Coping with Global Environmental Change, Disasters and Security—Threats, Challenges, Vulnerabilities and Risks* (Berlin–Heidelberg–New York: Springer-Verlag): 1487–1504.

Oswald Spring, Úrsula; Brauch, Hans Günter; Tidball, Keith G., 2014: "Expanding Peace Ecology: Peace, Security, Sustainability, Equity and Gender", in: Oswald Spring, Úrsula; Brauch, Hans Günter; Tidball, Keith G. (Eds.): *Expanding Peace Ecology: Peace, Security, Sustainability, Equity and Gender—Perspectives of IPRA's Ecology and Peace Commission* (Cham–Heidelberg–New York–Dordrecht–London: Springer): 1–32.

Randers, Jorgen, 2012: *2052—A Global Forecast for the Next 40 Years* (White River Jct., VT: Chelsea Green Publishing).

Raskin, P.; Banuri, T.; Gallopin, G.; Gutman, P.; Hammond, A.; Kates, A.; Swart, R., 2002: *Great transition. The promise and the lure of the times ahead.* Report of the Global Scdenario Group. Pole Start Series Report 10 (Stockholm: Stockholm Environment Institute).

Rotmans, Jan; Loorbach, Derk, 2010: "Towards a Better Understanding of Transitions and Their Governance: A Systemic and Reflexive approach", in: Grin, John; Rotmans, Jan; Schot, Johan (Eds.): *Transitions to Sustainable Development. New Directions in the Study of Long Term Transformative Change* (New York, NY–London: Routledge): 105–222.

Salehyan, Idean, 2014: "Climate change and conflict: making sense of disparate findings", in: *Political Geography*, 43,1: 1–5.

Salehyan, Idean; Hendrix, Cullen, 2014: "Climate shocks and political violence", in: *Global Environmental Change*, 28, 1: 239–250.

Scheffran, Jürgen; Brzoska, Michael; Brauch, Hans Günter; Link, Peter Michael; Schilling, Janpeter (Eds.), 2012: *Climate Change,Human Security and Violent Conflict: Challenges for Societal Stability* (Berlin–Heidelberg–New York: Springer-Verlag).

Schellnhuber, Hans Joachim; Cramer, Wolfgang; Nakicenovic, Nebojsa; Wigley, Tom; Yohe, Gary (Eds.), 2006: *Avoiding Dangerous Climate Change* (Cambridge: Cambridge University Press).

Steffen, Will; Sanderson, Angelina; Tyson, Peter D.; Jäger, Jill; Matson, Pamela A.; Moore III, Berrien; Oldfield, Frank; Richardson, Katherine; Schellnhuber, Hans Joachim; Turner II, B.L.; Wasson, Robert J., 2004: *Global Change and the Earth System. A Planet under Pressure. The IGBP Series* (Berlin–Heidelberg–New York: Springer-Verlag).

Steffen, Will; Grinevald, Jacques; Crutzen, Paul J.; McNeill, John, 2011: "The Anthropocene: Conceptual and Historical Perspectives", in: *Phil. Trans. R. Soc. A* 369: 843.

Stephenson, Carolyn, 2016: "Paradigm and Praxis Shifts: Transitions to Sustainable Environ-mental and Sustainable Peace Praxis", in: Brauch, Hans Günter; Oswald Spring, Úrsula; Grin, John; Scheffran; Jürgen (Eds.): *Handbook on Sustainability Transition and Sustainable Peace* (Cham–Heidelberg–New York–Dordrecht–London: Springer International Publishing).

Theisen, Ole Magnus; Gleditsch, Nils Petter; Buhaug, Halvard, 2013: "Is Climate Change a Driver of Armed Conflict?", in: *Climatic Change*, 117,3: 613–625.

UN, 2009: *Climate Change and Its Possible Security Implications. Report of the Secretary-General.* A/64/350 of 11 September 2009 (New York: United Nations).

UNEP, 2011: *Green Economy Report: Towards a Green Economy: Pathways to Sustainable Development and Poverty Eradication* (Nairobi: UNEP); at: http://www.unep.org/greeneconomy/GreenEconomyReport/tabid/29846.

UNEP, 2014: *Decoupling 2: Technologies, Opportunities and Policy Options- Report to UNEP's International Resource Panel* (Nairobi: UNEP).

UNGA, 2009: "Climate Change and its Possible Security Implications". Resolution Adopted by the General Assembly, A/RES/63/281 (New York: United Nations General Assembly, 11 June).

UNSC, 2007: "Security Council Holds First-Ever Debate on Impact of Climate Change on Peace, Security, Hearing over 50 Speakers, UN Security Council, 5663rd Meeting, 17 April 2007"; at: http://www.un.org/News/Press/docs/2007/sc9000.doc.htm.

UNSC, 2011: "Statement by the President of the Security Council on "Maintenance of Peace and Security: Impact of Climate Change", S/PRST/1011/15, 20 July 2011.

UNSG 2009: "Climate Change and its Possible Security Implications" of 11 September 2009 (A/64/350).

Von Weizsäcker, Ernst-Ulrich (Ed.) 2014: *Ernst Ulrich von Weizsäcker: A Pioneer on Environmental, Climate and Energy Policy*—Presented by Uwe Schneidewind. Springer Briefs on Pioneers in Science and Practice No. 28 (Cham–Heidelberg–New York–Dordrecht–London: Springer-Verlag).

Von Weizsäcker, Ernst; Lovins, Amory B.; Lovins, L. Hunter, 1997: *Factor Four: Doubling Wealth, Halving Resource Use* (London: Earthscan).

Von Weizsacker, Ernst; Hargroves, Karlson J.; Smith, Michael H.; Desha, Cheryl J. K.; Stasinopoulos, Peter, 2009: *Factor 5: Transforming the Global Economy through 80 % Improvements in Resource Productivity* (London: Earthscan).

Wæver, Ole, 1995: "Securitization and Desecuritization", in: Lipschutz, Ronnie D. (Ed.): *On Security* (New York: Columbia University Press): 46–86.

Wæver, Ole, 1997: *Concepts of Security* (Copenhagen: Department of Political Science).

WBGU, 2007: *Welt im Wandel—Sicherheitsrisiko Klimawandel* (Berlin–Heidelberg: Springer-Verlag).

WBGU, 2008: *World in Transition—Climate Change as a Security Risk* (London: Earthscan); at: http://www.wbgu.de/wbgu_jg2007_engl.html.

WBGU, 2011: *World in Transition—A Social Contract for Sustainability* (Berlin: German Advisory Council on Global Change, July 2011).

WCED (World Commission on Environment and Development), 1987: *Our Common Future. The World Commission on Environment and Development* (Oxford–New York: Oxford University Press).

International Peace Research Association (IPRA)

Founded in 1964, the International Peace Research Association (IPRA) developed from a conference organized by the "Quaker International Conferences and Seminars" in Clarens, Switzerland, 16–20 August 1963. The participants decided to hold international Conferences on Research on International Peace and Security (COROIPAS), which would be organized by a Continuing Committee similar to the Pugwash Conferences. Under the leadership of John Burton, the Continuing Committee met in London, 1–3 December 1964. At that time, they took steps to broaden the original concept of holding research conferences. The decision was made to form a professional association with the principal aim of increasing the quantity of research focused on world peace and ensuring its scientific quality.

An Executive Committee including Bert V.A. Roling, Secretary General (The Netherlands), John Burton (United Kingdom), Ljubivoje Acimovic (Yugoslavia), Jerzy Sawicki (Poland), and Johan Galtung (Norway) was appointed. This group was also designated as Nominating Committee for a 15-person Advisory Council to be elected at the first general conference of IPRA, to represent various regions, disciplines, and research interests in developing the work of the Association. Since then, IPRA has held 25 biennial general conferences, the venues of which were chosen with a view to reflecting the association's global scope. IPRA, the global network of peace researchers, has just held its 25th General Conference on the occasion of its 50th anniversary in Istanbul, Turkey in August 2014 where peace researchers from all parts of the world had the opportunity to exchange actionable knowledge on the conference broad theme of 'Uniting for sustainable peace and universal values'.

The 26th IPRA General Conference will take place between November 28 and 1st December in 2016 in Freetown, Sierra Leone on the theme: AGENDA FOR

H.G. Brauch et al. (eds.), *Addressing Global Environmental Challenges from a Peace Ecology Perspective*, The Anthropocene: Politik—Economics—Society—Science 4, DOI 10.1007/978-3-319-30990-3

PEACE AND DEVELOPMENT: Conflict prevention, post-conflict transformation, and the Conflict, Disaster and Development Debate.

On IPRA http://www.iprapeace.org/.
IPRA 2016 Conference Brochure: http://www.iprapeace.org/images/newsletters/IPRA%202016%20Freetown%20%20CONFERENCE%20%20BROCHURE.pdf.

On previous IPRA Conferences:
IPRA 2012 in Mie https://www.facebook.com/media/set/?set=a.321841277928978.77587.320866028026503&type=3.
IPRA 2014 in Istanbul https://www.facebook.com/ipra2014.

On the IPRA Foundation: http://iprafoundation.org/.

IPRA Conferences, Secretary Generals and Presidents 1964–2016

IPRA General Conferences	IPRA Secretary Generals/Presidents
1. Groningen, the Netherlands (1965)	1964–1971 Bert V.A. Roling (the Netherlands)
2. Tallberg, Sweden (1967)	1971–1975 Asbjorn Eide (Norway)
3. Karlovy Vary, Czechoslovakia (1969)	1975–1979 Raimo Väyrynen (Finland)
4. Bled, Yugoslavia (1971)	1979–1983 Yoshikazu Sakamoto (Japan)
5. Varanasi, India (1974)	1983–1987 Chadwick Alger (USA)
6. Turku, Finland (1975)	1987–1989 Clovis Brigagão (Brazil)
7. Oaxtepec, Mexico (1977)	1989–1991 Elise Bouding (USA)
8. Königstein, FRG (1979)	1991–1994 Paul Smoker (USA)
9. Orillia, Canada (1981)	1995–1997 Karlheinz Koppe (Germany)
10. Győr, Hungary (1983)	1997–2000 Bjørn Møller (Denmark)
11. Sussex, England (1986)	2000–2005 Katsuya Kodama (Japan)
12. Rio de Janeiro, Brazil (1988)	2005–2009 Luc Reychler (Belgium)
13. Groningen, the Netherlands (1990)	2009–2012 Jake Lynch (UK/Australia)
14. Kyoto, Japan (1992)	Katsuya Kodama (Japan)
15. Valletta, Malta (1994)	2012–2016 Nesrin Kenar (Turkey)
16. Brisbane, Australia (1996)	Ibrahim Shaw (Sierra Leone/UK)
17. Durban, South Africa (1998)	**Presidents**
18. Tampere, Finland (2000)	The first IPRA President was Kevin Clements
19. Suwon, Korea (2002)	(New Zealand/USA, 1994–98).
20. Sopron, Hungary (2004)	His successor was Úrsula Oswald Spring
21. Calgary, Canada (2006)	(Mexico, 1998–2000).
22. Leuven, Belgium (2008)	
23. Sydney, Australia (2010)	
24. Mie, Japan (2012)	
25. Istanbul, Turkey (2014)	
26. Freetown, Sierra Leone (2016)	

IPRA's Ecology and Peace Commission (EPC)

IPRA's *Ecology and Peace Commission* (EPC) addresses the relationship between the Earth and human systems, and their impacts on peace. A special focus is placed on the linkages between problems of sustainable development and sustainable peace. The EPC evolved from the Food Study Group, which became *Ecology and Peace Commission* (EPC). In 2004 an *Earth Charter Working Group* was also set up. Many wars have been related to resource conflicts and therefore the EPC focused on conflict resolution related to sustainable development and processes of sustainable transition toward ecological civilization.

The conveners are elected by the participants during IPRA conferences for a two year period to prepare the publications for the past conference and to prepare the sessions for the next conference. The conveners between the IPRA conferences in Mie (2012) and Istanbul (2014) were:

- Úrsula Oswald Spring (CRIM/UNAM, Cuernavaca, Mexico), Full time Professor/Researcher at the National University of Mexico (UNAM) in the Regional Multidisciplinary Research Center (CRIM), lead author of the Intergovernmental Panel on Climate Change (IPCC); Email: uoswald@gmail.com.
- Hans Guenter Brauch (Free University of Berlin (ret.), Peace Research and European Security Studies [AFES-PRESS], Mosbach, Germany); Chairman, Peace research and European Security Studies (AFES-PRESS), nonprofit scientific society, Mosbach, Germany; Email: brauch@afes-press.de;

H.G. Brauch et al. (eds.), *Addressing Global Environmental Challenges from a Peace Ecology Perspective*, The Anthropocene: Politik—Economics—Society—Science 4, DOI 10.1007/978-3-319-30990-3

- Keith G. Tidball (Cornell University, Ithaca. NY, USA), Senior Extension Associate in the Department of Natural Resources where he serves as Associate Director of the Civic Ecology Lab and Program Leader for the Nature & Human Security Program. New York State Coordinator for NY Extension Disaster Education Network; Email: kgtidball@cornell.edu.

Based on the presentations of the IPRA conference in Mie (November 2012) they published this peer-reviewed book:

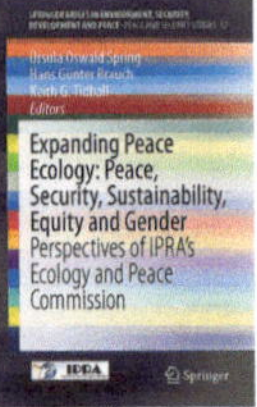

Úrsula Oswald Spring; Hans Günter Brauch; Keith G. Tidball (Eds.): *Expanding Peace Ecology: Security, Sustainability, Equity and Peace: Perspectives of IPRA's Ecology and Peace Commission 1.* SpringerBriefs in Environment, Security, Development and Peace, vol. 12. Peace and Security Studies No. 2 (Cham–Heidelberg–New York–Dordrecht–London: Springer-Verlag, 2014).
ISBN (Print): 978-3-319-00728-1
ISBN (Online/eBook): 978-3-319-00729-8
DOI: 10.1007/978-3-319-00729-8

In August 2014 in Istanbul the conveners between the IPRA conferences in Istanbul (2014) and in Freetown (2016) were elected:

- Prof. Dr. Úrsula Oswald Spring (CRIM/UNAM, Cuernavaca, Mexico)
- PD Dr. Hans Guenter Brauch (Free University of Berlin (ret.), Peace Research and European Security Studies [AFES-PRESS], Mosbach, Germany)
- Juliet Bennett, Ph.D. candidate (Centre for Peace and Conflict Studies, The University of Sydney Australia); Email: juliet.bennett@sydney.edu.au.

Based on the presentations of the IPRA conference in Istanbul (August 2014) they published these two peer-reviewed books:

- Hans Günter Brauch, Úrsula Oswald Spring, Juliet Bennett, Serena Eréndira Serrano Oswald (Eds.): *Addressing Global Environmental Challenges from a Peace Ecology Perspective.*
- Úrsula Oswald Spring, Hans Günter Brauch, Serena Eréndira Serrano Oswald, Juliet Bennett (Eds.): *Regional Ecological Challenges for Peace in Africa, the Middle East, Latin America and Asia Pacific.*

Mosbach, Germany
Cuernavaca, Mexico
Sydney, Australia
1 December 2015

Hans Günter Brauch
Úrsula Oswald Spring
Juliet Bennett

About the Editors

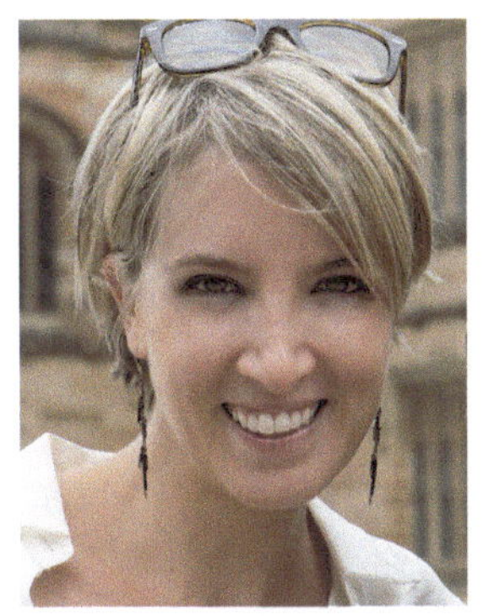

Juliet Bennett (Australia) is a Ph.D. Candidate at the Centre for Peace and Conflict Studies, the University of Sydney. Her research explores an interface between ecology, religion and peace. Juliet has presented her research at international conferences and published papers in academic journals and edited books. She has taught at Lenoir Rhyne University, North Carolina as well as at The University of Sydney. Juliet has been the Executive Officer of the Sydney Peace Foundation since 2012, and in 2014 she became a co-convener of the *Peace and Ecology Commission* (EPC) of the International Peace Research Association (IPRA).

Address: Centre for Peace and Conflict Studies, Mackie Building K01, The University of Sydney NSW 2006.
Email: juliet.bennett@sydney.edu.au.
Website: http://www.julietbennett.com.

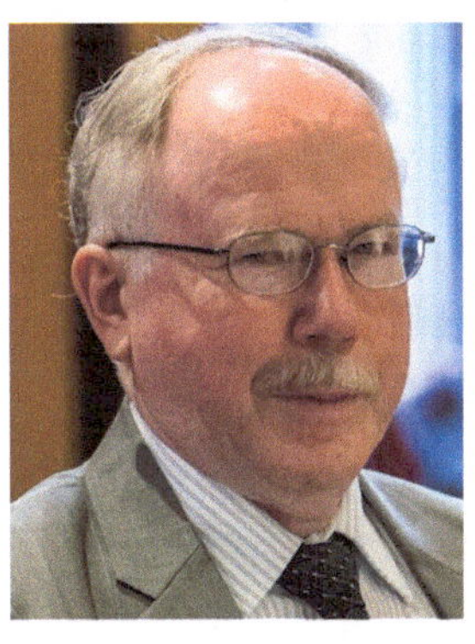

Hans Günter Brauch (Germany), Dr., Adj. Prof. (Privatdozent) at the Faculty of Political and Social Sciences, Free University of Berlin (ret.); since 1987 chairman of *Peace Research and European Security Studies* (AFES-PRESS). He is editor of the *Hexagon Book Series on Human and Environmental Security and Peace* (HESP), and of *SpringerBriefs in Environment, Security, Development and Peace* (ESDP), of the *SpringerBriefs of Pioneeres in Science and Practice, of the Pioneers in Arts, Humanities, Science, Engineering, and Practice* with Springer International Publishing. He was guest professor of international relations at the universities of Frankfurt on Main, Leipzig, Greifswald, and Erfurt; research associate at Heidelberg and Stuttgart universities, and a research fellow at Harvard and Stanford

H.G. Brauch et al. (eds.), *Addressing Global Environmental Challenges from a Peace Ecology Perspective*, The Anthropocene: Politik—Economics—Society—Science 4, DOI 10.1007/978-3-319-30990-3

Universities. In fall and winter 2013/2014 he was a guest professor at Chulanlongkorn University in Bangkok. He published on security, armament, climate, energy, and migration, and on Mediterranean issues in English and German, was translated into Spanish, Greek, French, Danish, Finnish, Russian, Japanese, Portuguese, Serbo-Croatian, and Turkish. Recent books in English: (co-ed. with Liotta, Marquina, Rogers, Selim): *Security and Environment in the Mediterranean. Conceptualising Security and Environmental Conflicts*, 2003; (co-ed. with Oswald Spring, Mesjasz, Grin, Dunay, Chadha Behera, Chourou, Kameri-Mbote, Liotta): *Globalization and Environmental Challenges: Reconceptualizing Security in the 21st Century, 2008;* (co-ed. with Oswald Spring, Grin, Mesjasz, Kameri-Mbote, Chadha Behera, Chourou, Krummenacher): *Facing Global Environmental Change: Environmental, Human, Energy, Food, Health and Water Security Concepts* (2009); (co-ed. with Oswald Spring): *Reconceptualizar la Seguridad en el Siglo XXI* (2009); (co-ed. with Oswald Spring, Mesjasz, Grin, Kameri-Mbote, Chourou, Dunay, Birkmann): *Coping with Global Environmental Change, Disasters and Security—Threats, Challenges, Vulnerabilities and Risks* (2011); (co-ed with Scheffran, Brzoska, Link, Schilling): *Climate Change, Human Security and Violent Conflict* (2012), and (co-ed with Oswald Spring, Grin, Scheffran): *Handbook on Sustainability Transition and Sustainable Peace* (2016).

Address: PD Dr. Hans Günter Brauch, Alte Bergsteige 47, 74821 Mosbach, Germany.
Email: brauch@afes-press.de.
Website: http://www.afes-press.de and http://www.afes-press-books.de/.

Úrsula Oswald Spring (Mexico), full time Professor/Researcher at the National University of Mexico (UNAM) in the Regional Multidisciplinary Research Center (CRIM), she was national coordinator of water research for the National Council of Science and Technology (RETAC-CONACYT), first Chair on Social Vulnerability at the United National University Institute for Environment and Human Security (UNU-EHS); founding Secretary-General of El Colegio de Tlaxcala; General Attorney of Ecology in the State of Morelos (1992–1994), National Delegate of the Federal General Attorney of Environment (1994–1995); Minister of Ecological Development in the State of Morelos (1994–1998). She was President of the International Peace Research Association (IPRA, 1998–2000), and General Secretary of the Latin-American Council for Peace Research (2002–2006). She studied medicine, clinical psychology, anthropology, ecology, classical and modern languages. She obtained her Ph.D. from University of Zürich (1978). For her scientific work she received the Price Sor Juana Inés de la Cruz (2005), the Environmental Merit in Tlaxcala, Mexico (2005, 2006), UN Development Prize.

She was recognized as Women Academic in UNAM (1990 and 2000); and Women of the Year (2000). She works on non-violence and sustainable agriculture with groups of peasants and women and is President of the Advisory Council of the Peasant University. She has written 46 books and more than 328 scientific articles and book chapters on sustainability, water, gender, development, poverty, drug consumption, brain damage due to under-nourishment, peasantry, social vulnerability, genetic modified organisms, bioethics, on human, gender, and environmental security, peace and conflict resolution, democracy, and conflict negotiation.

Address: Prof. Dr. Úrsula Oswald Spring, CRIM-UNAM, Av. Universidad s/n, Circuito 2, Col. Chamilpa, Cuernavaca, CP 62210, Mor., Mexico.
Email: uoswald@gmail.com.
Website: http://www.afes-press.de/html/download_oswald.html.

Serena Eréndira Serrano Oswald (Mexico) is research professor at the Regional Multidisciplinary Research Centre, National Autonomous University of Mexico (CRIM-UNAM). She holds a Ph.D. in Social Anthropology (UNAM), an MSc in Social Psychology (LSE), an MFT in Systemic Family Therapy (CRISOL), and a BA Hons in Political Studies and History (SOAS). She has a Postdoctorate in Sociology and Gender (UNAM), a professional diploma in translation and interpreting (Institute of Linguists), a specialized training in couples therapy, in psychopathology (CRISOL), and person-centred therapy (Gestalt Institute). Certified by the National Council of Researchers (SNI I), she is currently president of the Mexican Association of Regional Development (AMECIDER).

Address: Dr. Serena Eréndira Serrano Oswald, Priv. Rio Bravo No. 1, Col. Vista Hermosa, Cuernavaca, Morelos, Mexico CP 62290.
Email: sesohi@gmail.com.

About the Contributors

Katharina Bitzker (Germany/Canada) is a medical doctor, body psychotherapist, peace researcher and writer. She received her medical doctoral degree from Humboldt-University Berlin (Germany) and also holds a MA in International Peace Studies from the UN-mandated University for Peace in Costa Rica. Her current areas of interest and research include the interconnectedness of love and peace, the role of poetry, music and humour in conflict transformation, systems theory approaches, and the convergence of neurobiology and peace studies. She is currently pursuing her Ph.D. in Peace and Conflict Studies at the University of Manitoba (Canada).

Address: Dr. Katharina Bitzker, Mauro Centre for Peace and Justice, University of Manitoba, 70 Dysart Road, Winnipeg, MB, R3T 2N2, Canada.
Email: katharina.bitzker@hotmail.de.
Website: http://umanitoba.ca/mauro_centre/.

Henri Myrttinen (Finland/UK) is the Head of the Gender Team of International Alert, a London-based peacebuilding organization. He has been researching, working, teaching, conducting trainings and publishing on gender, especially in the context of post-conflict and conflict-affected societies with and for a range of stakeholders, beneficiaries and target audience for the past 15 years. He holds a Ph.D. in Conflict Resolution and Peace Studies from the University of KwaZulu-Natal, South Africa, with a thesis on masculinities and violence. Other research on gender issues includes examining security sector reform issues, small arms violence, the reintegration of ex-combatants and displaced populations, and prevention of and responses to sexual and gender-based violence from a comprehensive gender perspective. He is the co-author of "Sexed Pistols—Gendered Impacts of Small Arms and Light Weapons" (UNU Press, 2009, with Vanessa Farr and Albrecht Schnabel).

Address: Dr. Henri Myrttinen, International Alert, 346 Clapham Road, London SW9 9AP, UK.
Email: hmyrttinen@international-alert.org.
Website: http://www.international-alert.org/gender.

H.G. Brauch et al. (eds.), *Addressing Global Environmental Challenges from a Peace Ecology Perspective*, The Anthropocene: Politik—Economics—Society—Science 4, DOI 10.1007/978-3-319-30990-3

About this Book

Addressing Global Environmental Challenges from a Peace Ecology Perspective offers peer-reviewed texts that build on *Expanding Peace Ecology* and applies this concept to *global environmental challenges* in the Anthropocene. *Hans Günter Brauch* (Germany) offers a typology of time and turning points in the 20th century; *Juliet Bennett* (Australia) discusses the global ecological crisis as resulting from a "tyranny of small decisions". *Katharina Bitzker* (Canada) debates "The Emotional Dimensions of Ecological Peacebuilding" by loving nature. *Henri Myrttinen* (UK) analyses "Preliminary findings on gender, peacebuilding and climate change in Honduras". *Úrsula Oswald Spring* (Mexíco) offers a critical review of the policy and scientific nexus debate on "The Water, Energy, Food and Biodiversity Nexus" reflecting on the case of security in Mexico. In closing, *Brauch* discusses whether strategies of sustainability transition may enhance the prospects for achieving sustainable peace in the Anthropocene.

- Addresses global environmental challenges
- Focuses on the nexus among biodiversity, water, food, energy and waste
- Deals with structural violence, the tyranny of small decisions and emotional dimensions of ecological peacebuilding
- Offers perspectives on sustainable peace by moving towards sustainability transition

Foreword by Nesrin Kenar, Sec. General, IPRA
Introduction (Hans Günter Brauch–Úrsula Oswald Spring—Juliet Bennett and Serena Eréndira Serrano Oswald)—2 Historical Times and Turning Points in a Turbulent Century: 1914, 1945, 1989 and 2014? (Hans Günter Brauch)—3 Global Ecological Crisis: Structural violence and the tyranny of small decisions (Juliett Bennett)—4 Loving Nature: The Emotional Dimensions of Ecological Peacebuilding (Katharina Bitzker)—5 Drowning in complexity? Preliminary findings on addressing gender, peacebuilding and climate change in Honduras (Henri Myrttinen)—6 The Water, Energy, Food and Biodiversity Nexus: New Security

H.G. Brauch et al. (eds.), *Addressing Global Environmental Challenges from a Peace Ecology Perspective*, The Anthropocene: Politik—Economics—Society—Science 4, DOI 10.1007/978-3-319-30990-3

Issues in the Case of Mexico (Úrsula Oswald Spring)—7 Building Sustainable Peace by Moving Towards Sustainability Transition (Hans Günter Brauch).

Backmatter: IPRA and on EPC—About the Editors and Contributors—About this Book.

More on this book at: http://www.afes-press-books.de/html/APESS_04-05.htm.